D1117379

Nuclear Reactions

Nuclear Reactions

THE POLITICS OF OPENING

A RADIOACTIVE

WASTE DISPOSAL SITE

Chuck McCutcheon

UNIVERSITY OF NEW MEXICO PRESS

ALBUQUERQUE

In memory of Dave Holden

and Susan Landon

© 2002 by the University of New Mexico Press
All rights reserved.
First Edition

LIBRARY OF CONGRESS CATALOGING-IN-PUBLICATION DATA

McCutcheon, Chuck, 1963–
Nuclear reactions : thte politics of opening a radioactive
waste disposal site / Chuck McCutcheon.—1st ed.
p. m.
Includes index.
ISBN 0-8263-2209-3 (cloth)
1. Waste Isolation Pilot Plant (N.M.) 2. Radioactive waste disposal in the ground—
New Mexico—Carlsbad Region. 3. Radioactive waste disposal—Political aspects—
New Mexico—Carlsbad Region. I. Title.
TD898.12.N6 M38 2002
363.72'89'0973—dc21
2002006711

DESIGN: Mina Yamashita

Contents

Acknowledgments

This book is a product of more than a dozen years of reporting on nuclear energy. Such an undertaking has enabled me to talk to, work with, and learn from a number of helpful people in the political, legal, scientific, and journalistic fields. The list includes but is not limited to Bruce Daniels, Isabel Sanchez, Dave Staats, John Fleck, Richard Parker, Tony Davis, Karen MacPherson, Bob Neill, Lokesh Chaturvedi, Matthew Silva, Wendell Weart, Don Hancock, Melinda Kassen, Roger Anderson, Lindsay Lovejoy, Margret Carde, Tom Udall, Jeff Bingaman, Pete Domenici, Bill Richardson, Joe Skeen, John Arthur, Bob Forrest, Chris Wentz, Chris Whipple, Skip Gosling, Rich Marquez, Beth Farrell, Tarek Khreis, Jim Bickel, and Tracy Loughead.

When I began thinking about a book on WIPP, I was fortunate to meet David Holtby of the University of New Mexico Press, whose unflagging enthusiasm and devotion to making it a reality have been inspiring. Thanks also are due to Michael Gerrard for his encouragement, as well as Luther Carter, Tony Rosenbaum, and Joe La Grone for their willingness to read the manuscript and offer suggestions, and Rob Dean and Dan Balduini for providing photographs. And Judi Hasson deserves kudos for her insightful editing.

Chronology of Waste Isolation Pilot Plant (WIPP)-Related Events

1957 The National Academy of Sciences concludes in a report commissioned by the U.S. Atomic Energy Commission (AEC) that "the most promising method" of disposing of radioactive wastes appears to be in underground salt deposits.

1972 After nearly a decade of study, an underground salt mine near Lyons, Kansas, is judged unacceptable after Kansas officials raise technical concerns. Atomic Energy Commission officials announce they will examine southeastern New Mexico as a potential waste storage site after being invited by a group of Carlsbad leaders. Officials hope to have a pilot facility ready by 1980.

1974 A location 30 miles east of Carlsbad is chosen for exploratory work and field investigations.

1979 Congress authorizes WIPP as "a research and development facility to demonstrate the safe disposal of radioactive wastes resulting from defense activities."

1981 New Mexico attorney general Jeff Bingaman files suit in U.S. District Court alleging violations of federal and state laws in connection with WIPP's development. The state of New Mexico subsequently reaches agreement with the Department of Energy (DOE) on a formal role for the state.

1988 Delays in WIPP's opening prompt Idaho governor Cecil Andrus to impose a ban on out-of-state waste shipments to the Idaho National Engineering Laboratory. Andrus subsequently agrees to allow shipments to resume temporarily while the department tries to open WIPP.

1989   Energy Secretary James Watkins announces an indefinite delay in WIPP's opening, saying the project "will only open when I deem it safe and other key non-DOE reviewers are satisfied."

1991   Watkins notifies Interior Secretary Manuel Lujan Jr. in October that WIPP is ready to begin accepting wastes for a five-year test phase. New Mexico attorney general Tom Udall and a coalition that includes the state of Texas and several environmental groups file a lawsuit asking that the DOE first receive congressional approval before wastes are sent to WIPP. In November, U.S. District Judge John Garrett Penn issues an order granting the state's motion for a preliminary injunction.

1992   President George Bush signs into law a bill establishing prerequisites for initial receipt and permanent disposal of waste at WIPP. The bill designates the Environmental Protection Agency (EPA) as WIPP's independent regulator.

1993   Energy Secretary Hazel O'Leary announces that tests with waste will be conducted in laboratories rather than at WIPP in the wake of continued criticism that the underground tests would be of little scientific merit.

1996   Congress passes legislation to amend the 1992 WIPP law by removing and changing several of the criteria in the law.

1997   The EPA issues a proposed rule certifying that WIPP complies with its disposal standards.

1998   The EPA announces its certification of WIPP's compliance with the standards. Energy Secretary Federico Peña notifies Congress that WIPP is ready. Peña later notifies the state of New Mexico that the DOE intends to ship waste without receiving a permit regulating nonmixed wastes under the Resource Conservation and Recovery Act (RCRA).

1999 After negotiations between the DOE and New Mexico fail to produce an agreement, Energy Secretary Bill Richardson sends the first shipments of waste to WIPP from Los Alamos National Laboratory in March. Shipments follow from Idaho National Engineering and Environmental Laboratory, Colorado's Rocky Flats plant, and other sites.

2002 WIPP receives its five-hundredth shipment in January.

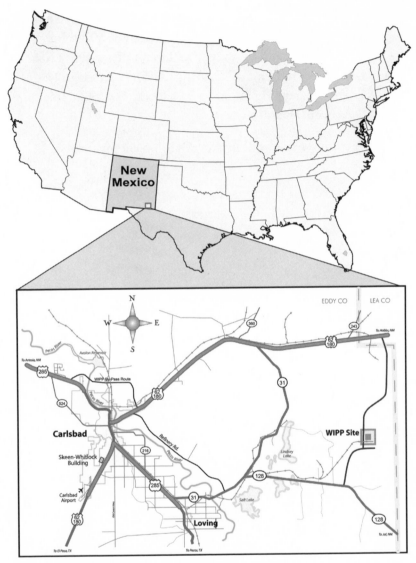

Figure 1: *Location of WIPP (Courtesy of Department of Energy).*

Figure 2: *(facing page) Stratigraphy of the underground area at WIPP Courtesy of Department of Energy).*

Introduction

The Waste Isolation Pilot Plant (WIPP), situated beneath the desolate and remote flatlands of southeastern New Mexico, is the first facility of its kind in the world—the most elaborate landfill built to date to permanently house mankind's deadliest garbage. Open since 1999, the project is intended to entomb, in underground rock salt, four decades' worth of protective gloves, tools, cleaning rags, glassware, and other cast-off items that have been contaminated with radioactive materials used in building nuclear weapons. It cost around $2 billion to build and $200 million a year to operate; the final price tag for preparing, shipping, and disposing of waste there has been estimated to be as high as $29 billion.[1] More than 6 million cubic feet of waste will be sent to the site—an amount equal to 850,000 55-gallon drums, enough to fill more than 65 rooms that are each about the size of a football field.

The underground area of WIPP where the waste is buried looks like

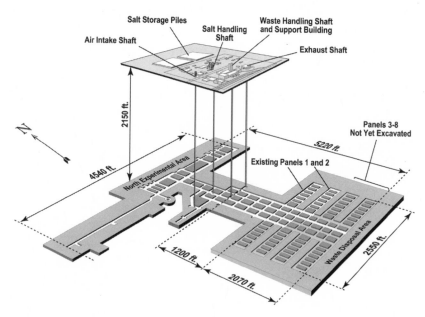

no other mine. Situated more than 2,000 feet inside the earth and accessible by a five-minute elevator ride, it features a vast seven-mile network of man-made tunnels connecting the cavernous storage rooms. Measuring gauges and monitoring instruments line the floors, walls, and ceilings, which are flecked with gray and orange-pink salt crystals. The circulation of air within the facility is strictly controlled, but dust gets around; workers say that after enough time underground, they can taste the gritty salt for days afterward. "The place," observed science fiction author Gregory Benford, "resembles a sort of subterranean, Borgesian, infinite parking garage."[2]

The materials heading to WIPP travel by truck from more than a dozen federal bomb factories and research complexes. They include the former Rocky Flats plant near Denver, New Mexico's Los Alamos National Laboratory, Idaho National Engineering and Environmental Laboratory, Washington State's Hanford Nuclear Reservation, Tennessee's Oak Ridge National Laboratory, and South Carolina's Savannah River Plant. The wastes are often described as "low-level," but that characterization is mis-leading. Instead they are "transuranic," meaning they contain elements heavier than uranium. Transuranic waste is neither high-level nor low-level, although it shares characteristics of both. The main transuranic element in WIPP's waste is plutonium, a man-made element used in making the "trigger" that sets off a nuclear warhead. Plutonium's alpha particles travel only inches and cannot penetrate skin or even a sheet of paper. Workers can actually handle most waste drums by hand, as long as they remain sealed. But if it is inhaled, swallowed, or absorbed into the bloodstream through a cut, an amount of plutonium as small as one-millionth of a gram can cause cancer. Plutonium not only is dangerous but long-lived—it takes 24,360 years to lose half of its radioactivity.

The theory behind WIPP is that after the wastes are put 2,150 feet underground, the rock salt will naturally collapse around them, forming a tight cocoon that seals them off and prevents their escape. Salt has been shown to move rapidly to heal fissures and close openings, and it is expected that less than a century after the final waste drum is buried, the materials will be completely encapsulated. Aboveground, meanwhile, large and elaborately designed markers will serve as futuristic "no trespassing" warnings. They will contain a variety of symbols in languages other than English, since there is no guarantee that modern words will outlive the dangerousness of the waste. The expectation is that no one will drill or dig into the site and allow radioactivity to escape and that no natural geologic occurrences will allow the materials to reach groundwater supplies or nearby rivers.

Such a concept has won widespread endorsement, and similar subterranean systems have been explored in other countries such as Sweden and Germany. Among scientists and policy makers, the general belief is that deep geologic disposal is preferable to other alternatives that have been considered, including burying wastes beneath the ocean floor or shooting them into outer space. A panel of international experts concluded in 1997 that the complex process based on computer modeling to judge WIPP's long-term safety was "in the main, technically sound."[3] Nevertheless, the New Mexico project has been one of the most tangled energy policy quagmires of recent decades. Before the initial waste shipment in March 1999, it had been actively planned for a quarter of a century, and in essence built for years. But it was plagued by more than a decade of delays and has remained controversial since its opening. The problems and fights that led to those delays and fueled those controversies have widened the gulf between America's ability to create sophisticated weapons and its ability to successfully show it can clean up after them.

At WIPP and elsewhere, the issue of how best to dispose of the Cold War's toxic trash has been a primary responsibility of the U.S. Department of Energy (DOE). Disposing of such dangerous materials requires that the department meet several challenges: transporting them to a remote location, keeping them isolated without leaks, spills, or other accidents, and demonstrating they can be adequately sealed off from society for thousands of years. The risks are heightened by the fact that waste storage affects many citizens. At the time of WIPP's initial shipment, 61 million people—roughly one-quarter of the nation's population—lived within 50 miles of a storage site for the military's nuclear waste. Trucks heading to New Mexico are using highways in 22 of the 48 continental states. (Although the department temporarily suspended shipments for national security reasons after the September 2001 terrorist attacks in New York and Washington, the trucks are not considered vulnerable to hijackers because the plutonium particles in the wastes

Figure 3: *Aerial view of the WIPP site (Department of Energy photo).*

cannot be converted into materials pure enough to be used in nuclear weapons.)[4]

The biggest challenge in storing nuclear waste has been overcoming public resistance. Despite technological advances that have increased confidence among scientists that wastes can be disposed of safely, the government and nuclear industry have fared poorly in helping the public overcome the fear and anxiety associated with storing these wastes. Studies of risk perception have shown that nuclear reactors and wastes trigger a reaction of dread; a repository has been judged to pose risks at least as great as a nuclear power plant or nuclear weapons test site. Those fears have been traced, in part, to the historical evolution of common perceptions that radioactivity is linked, in many people's minds, with death and disaster. As one study noted: "Nuclear energy was conceived in secrecy, born in war and first revealed to the world in horror. No matter how much proponents try to separate the peaceful from the weapons atom, the connection is firmly embedded in the minds of the public."[5]

Obviously, then, radioactive waste is fully capable of causing a furor on

Figure 4: Cutaway view of tools, glassware, clothing, and other wastes inside drums
to be buried at WIPP (Department of Energy photo).

a par with other divisive hot-button issues. "With the possible exception
of the income tax," journalists Donald Barlett and James Steele observed,
"no other modern-day issue is so firmly mired in Washington politics as
that of nuclear waste."[6] With WIPP, environmentalists and other critics
have vociferously argued that the plant is an ill-advised attempt to
address waste storage. They contend the site is inherently unsuitable, given
its location in an area known for abundant oil and gas resources. They have
raised a litany of concerns: pockets of pressurized salt water that could
carry radioactivity to the surface, water seepage that could corrode waste
drums and result in waste getting outside the site, underground cracks
and ceiling collapses that have occurred at a faster rate than anticipated.
In addition to technical arguments, the critics have contended that the
way the site was picked and proved safe was riddled with flaws. Political
expediency, they say, repeatedly triumphed over technical merit and
careful external scrutiny. "Over and over again, the history of WIPP shows
a democratic process that has been short-circuited and a scientific
process that has been ignored," one anti-WIPP organization charged.

"If regulations or scientific criteria create problems, they are changed or deleted. If promises are inconvenient, they are broken."[7]

Critics argue that the plant will not "solve" the government's nuclear waste storage problem because it remains relatively small in scope: it is slated to dispose of less than one-quarter of the wastes generated from weapons manufacturing. Yet these critics say so much money and attention has been devoted to opening and running the site that it has distracted the government from tackling larger waste management issues. Predictably, fears about it have given rise to the "not in my backyard" (NIMBY) arguments prevalent in many communities confronted with providing a home to nuclear, toxic, or hazardous materials. The NIMBY problem has spawned related acronyms characterizing deep-seated public opposition to siting processes, including BANANA ("build absolutely nothing anywhere near anything").[8] Because of its unique circumstances, however, WIPP illustrates how many separate and broader challenges await future storage sites: the role and elusive nature of consensus in shaping public opinion, the complexity of resolving federal-state government conflicts, and the limits of scientific truth. In that regard, it can be seen as a prototype of the ambiguous and problematic environmental issues facing the United States and other nations in the twenty-first century.

A History of Procrastination

Many of these challenges arose because the United States did not look seriously at what to do with radioactive waste until years after it began producing such waste—a situation that has been compared to building a house without any toilets.[9] After the bombs dropped on Japan ended World War II, scientists and policy makers hoped to find an atomic solution to the nation's problems. Eager politicians envisioned a future in which ships, trains, and cars all would run on nuclear energy before the year 2000. The Atomic Energy Commission (AEC), established in 1947, put its emphasis on increasing supplies of uranium ore and enriched uranium, producing plutonium, and building more weapons. Managing the waste was a distant concern. The thinking was that if the United States could develop the technology to harness the awesome power of the atom, it most assuredly could deal with disposing of its by-products—someday.[10]

Among those who regarded waste management as relatively unimportant was J. Robert Oppenheimer, the brilliant scientist who led the Manhattan Project, which developed the nuclear bomb at Los Alamos. That attitude was reflected in the laboratory's practice during the late 1940s of simply dumping radioactive and toxic materials into adjacent canyons.[11] Similar attitudes prevailed elsewhere. For 15 years beginning in 1954, trucks and trains hauled waste from Colorado's Rocky Flats plant to Idaho, where the materials were packed in cardboard boxes or steel barrels and buried in shallow, unlined, dirt-covered pits. "It was easier to just throw them away than to check them for radioactivity and decontaminate them," said Bruce Schmalz, the former head of waste management at Idaho National Engineering Laboratory. If anything was suspected of being contaminated, he said, "they just simply dumped it into a trench and covered it up."[12]

Much of the waste at the Idaho lab—including transuranic materials—was classified for many years as "low-level," a broad and somewhat vague category. It is not high-level waste: the highly radioactive liquids and other materials resulting from chemical reprocessing of reactor fuel to recover usable uranium and plutonium. In reprocessing, uranium ore is crushed, ground, and chemically processed to produce uranium oxide, known as "yellowcake" because of its bright color. It is then converted to uranium hexafluoride, which becomes a gas that permits the enrichment of uranium 235, the isotope required for reactor fuel. The enriched gas is converted into solid uranium oxide, which is formed into ceramic pellets to make fuel rods. The rods are then bundled into fuel assemblies containing 50 to 300 rods. Rods that no longer contribute to a nuclear chain reaction are known as "spent fuel."

Low-level waste, meanwhile, comes from the core and cooling systems of nuclear reactors, while some comes from the tools, clothing, glassware, and other debris used in weapons work. Commercial low-level waste comes from medical and academic research. The Low-Level Waste Act, signed into law in 1980, made states responsible for disposing of such wastes within their borders. Although most low-level waste is relatively short-lived, some of it is "hot" enough to require special protective shielding for handling and shipment.[13]

During the 1950s, radioactive by-products of all kinds accumulated as weapons production plants sprang up across the country in the wake of Soviet Russia's detonation of its first atomic bomb in 1949. On the civilian end, Congress passed the Atomic Energy Act of 1954, allowing the federal government and private industry to become partners to promote nuclear power. But the military side remained dominant: weapons research and production came to yield four times as much waste as commercial nuclear power plants. In both instances, the attention remained on production, with few top scientists drawn to studying storage. "There were no careers; it was messy," recalled Carroll Wilson, the AEC's first general manager. "Nobody got brownie points for caring about nuclear waste."[14] The apathy translated into little money—the agency's entire budget for waste management in 1961 was $9.2 million.[15] Nevertheless, commission officials stated that waste management operations were "more satisfactory than essentially any other facet of the nuclear fuel cycle."[16]

Besides being optimistic, federal officials were notoriously secretive. The evolution of such a closed-off management approach has been another reason why the successful storage of nuclear waste has proved so difficult to sell to the public. Within the nuclear weapons establishment during most of the twentieth century, winning the Cold War took priority over informing citizens. Government documents were routinely classified, journalists were given misleading information, and the public was fed propaganda, all in the name of national security. When nuclear bombs were exploded in the Marshall Islands in the Pacific in the late 1950s, officials assured the public that the radiation exposure was no worse than from receiving a chest x ray. Such claims were disproved after a Japanese fishing trawler met a radioactive cloud, causing acute radiation sickness of all on board and killing one sailor by the time the boat reached home two weeks later.[17]

Despite the public fears triggered by such incidents, potential hazards at weapons plants were brushed off as inconsequential, and most workers downplayed the risks associated with their jobs. Some scholars believe the location of many of the facilities in remote areas in the West— particularly New Mexico, Nevada, Idaho, and Washington—helped contribute to such a mind-set. "The folkways of Western nuclear culture parallel the folkways of other Western work cultures—cowboys, miners,

Figure 5: *Workers at Los Alamos National Laboratory load waste into containers for shipment to WIPP (Los Alamos National Laboratory photo).*

loggers, railroad workers—in a tough dismissal of danger," said University of Colorado history professor Patricia Nelson Limerick.[18] For decades, then, the public was largely inclined to support existing policy. In 1960, a nuclear plant siting survey in New York asked residents if they felt confident about the AEC's disposal safety rules. Fifty-seven percent said they were confident, 13 percent had some question, and 30 percent didn't know.[19]

However, the emergence of severe and widespread environmental and health problems at many sites eventually caused considerable alarm and anger. The scope of cleaning up the Energy Department's polluted weapons complex has become staggering: the agency estimates addressing the risks from five decades of production at around 100 sites will cost at least $150 billion to $200 billion—and take 70 years to complete. That cost comes on top of the more than $50 billion that already has been spent since 1989. Even after all the money is spent, long-term stewardship will be needed to ensure the public does not come into contact with the hazards that will remain.[20] Besides the environmental problems at weapons plants, incidents such as the 1979 Three Mile Island mishap in Pennsylvania and the 1986 Chernobyl reactor explosion in the former Soviet Union have created more mistrust of nuclear energy than of any other energy source. The result has been that waste cleanup efforts since the 1970s have been held to a far higher standard than any prior projects. As mistrust and skepticism have deepened, new projects—most notably WIPP—have had to clear a bar of public and political acceptance that has been set much higher than it would have been a few decades ago.

Such skepticism flourished as society became increasingly disillusioned with government during the 1960s and 1970s. Before then, discussions about nuclear waste took place on a scientific rather than sociological level. The AEC set up a committee of geologists and geophysicists in 1955 under the auspices of the National Academy of Sciences to study underground nuclear waste burial. Two years later, the academy issued a study concluding that "the most promising method of disposal" appeared to be in underground salt. "The great advantage here is that no water can pass through salt," it said. "Fractures are self-sealing. Abandoned salt mines or cavities especially mined to hold waste are, in essence, long-enduring tanks." The report provided a guidepost for the government to look at specific states. As it noted, salt beds and mines were abundant along the south side of the Great Lakes from New York to Michigan, in "salt domes" along much of the Gulf Coast, and large deposits elsewhere.[21]

From Lyons to Carlsbad

The Carey mine, near Lyons, Kansas, in one of the deposits, became the focus of the government's geologic disposal efforts during the 1960s. The study was part of a program called Project Salt Vault, which called for used fuel rods from commercial reactors to be reprocessed and the leftover liquid waste solidified and buried underground. In 1965, metal canisters with used fuel from a test reactor in Idaho were placed in 12-foot-deep holes drilled in the floor of the mine and monitored for 18 months. Although cracks and corrosion developed in the steel walls of the canisters, officials considered the project a success. By 1970, the commission tentatively designated the Carey mine as the first U.S. repository "if the salt meets design and geological criteria." Congress gave the commission $3.5 million in 1971 to buy the mine, acquire land around it, and prepare a conceptual design of the repository, which Commissioner James T. Ramey boasted would "last for centuries."[22]

As it happened, the impetus to use the mine came from political pressures. The General Accounting Office, the investigative arm of Congress, reviewed high-level-waste activities in 1968 and called on the AEC to devote greater attention and resources to the problem. However, its recommendations were largely ignored. Then on May 11, 1969, a massive fire struck Rocky Flats, causing more than $70 million in damages. Monitoring after the fire turned up evidence of previous contamination from a 1957 fire and prompted environmentalists to press for more information. The AEC eventually acknowledged the presence of airborne off-site contamination and plutonium-contaminated wastes buried both inside and outside the plant's gates. The fire prompted officials to relocate 330,000 cubic feet of waste to Idaho. When the media reported on the issue, the state's politicians reacted with outrage and pressured the federal government to begin looking seriously at underground disposal.[23]

As scientists began studying the Kansas mine, the AEC promoted its cause by estimating that the project would hire 200 employees and attract other industries. But Kansas's state government leaders conducted an investigation that led to several troubling discoveries. They found the land above the mine had been heavily drilled for oil and gas, and some boreholes had not been properly plugged. Kansas Geological Survey director William Hambleton described the ground as "a bit like a piece of Swiss cheese." If water penetrated the boreholes and seeped into the mine, radioactive brine could reach groundwater supplies. The federal government's entire approach to studying the issue also irked the Kansans. They complained that the Oak Ridge National Laboratory scientists overseeing the project were not addressing how heat from high-level waste would affect the surrounding salt and rock. "It has seemed to us at times that the AEC has been more interested in convincing the public of the safety of the Lyons site rather than using these funds needed to carry studies to a conclusion," Hambleton said in a letter to Governor Robert B. Docking.[24] The scientific findings touched off protests and congressional criticism, and by 1972, the repository was declared dead.

It was after they learned about the problems and public opposition in Kansas that a group of ambitious residents in Carlsbad, New Mexico, became interested in hosting their own waste storage facility. They persuaded state officials to go along, lobbied the federal government to look at their region, and, in time, a site was chosen and a repository built 26 miles east of the city. Carlsbad's local power structure has consistently supported the project since then, offering an interesting twist on the NIMBY syndrome. The plant, which employs around 800 people on-site, has given the city of 27,000 the financial stability its leaders desired: the Energy Department estimated that it contributed $161 million to the local economy in 1998.[25] But the federal government did not learn from its mistakes at Lyons in underestimating the powers of state government opposition. Although New Mexico officials accepted studying the project, they eventually sought veto power over it. Their concerns were rebuffed, causing a rift in federal-state relations that led to a lawsuit giving the state a formal oversight role. The lawsuit, however, did little to end the intergovernmental conflicts, which have continued past WIPP's opening.

The federal government did not anticipate other obstacles:

● Congress supplanted the executive branch as the major driver in the nuclear waste debate in the late 1970s and early 1980s. Lawmakers became involved in a controversy over whether commercial waste should be buried at the plant, only to have President Jimmy Carter cancel WIPP entirely. The dispute ended with the House and Senate overriding Carter to dictate the terms of its use.

● In the 1980s, the project became subject to regional political and environmental pressures. Western governors discovered they could take advantage of public fears about nuclear materials to lobby the Energy Department to use WIPP before some critics believed all scientific and bureaucratic hurdles had been cleared. At the same time, environmentalists who opposed nuclear power focused on the waste storage issue and were able to mobilize opposition based on scientific disputes that were not always easily resolvable or clear-cut, creating a climate of considerable uncertainty.

- The formal method developed in the 1990s to assess the crucial question of the plant's safety was slow in coming and did not settle the issue to everyone's satisfaction. Congress wrestled with the question of how to adequately judge safety, eventually setting up a bureaucratic process that put the Environmental Protection Agency (EPA) and Energy Department in conflict with each other. The EPA's decision to certify the site as safe was met with skepticism by many critics.

- The judiciary ultimately replaced Congress as the determining force for WIPP. A federal judge put plans to open the project on hold by ruling that Congress needed to pass legislation first. Later, the same judge granted a motion that overrode the state government's objections and allowed shipments to start.

The result of all these developments was that it took almost 20 years longer for WIPP to open than initially planned. The debate dragged on so long that it became difficult to sustain any continuity among governmental agencies—and practically impossible to muster consensus between supporters and opponents.

The facility's ability, in the end, to surmount such obstacles makes it a unique case study in U.S. environmental policy. The resulting perch it occupies in modern New Mexico culture makes it all the more interesting. The state's news media, as well as the national press, has relentlessly chronicled its development, making it familiar to most New Mexicans. It has become the stuff of fiction: Rudolfo Anaya's 1992 mystery *Zia Summer* involves a cunning activist who tries to sabotage a waste shipment, while Jake Page's 2000 thriller *Cavern* deals with the mysterious disappearance of WIPP workers. In the progressive arts colony of Santa Fe, artists have incorporated WIPP into sculptures and other works to protest its use. Meanwhile in Carlsbad, which is home to another famous underground attraction, Carlsbad Caverns National Park, vacationing tourists can stroll through the Energy Department's visitor center to look at an exhibit featuring drums filled with mock radioactive materials. Scientists have found the site fascinating for other reasons: in October 2000, biologists examined salt dug from WIPP's construction and found bacteria that they estimated at 250 million years old—10 times older than the oldest-known living organism.[26]

With more attention being paid to cleaning up nuclear materials, replays of the WIPP experience are likely. Some of the conflicts, in fact, have surfaced in the United States' other major nuclear waste undertaking: the much larger Yucca Mountain project in Nevada, aimed at burying spent fuel largely from commercial nuclear power plants. The federal government hopes to ready Yucca Mountain, 100 miles northwest of Las Vegas, to accept wastes from more than 80 reactor sites. Initially, storage facilities for spent fuel were created in the 1950s and 1960s with the assumption they would be stored in underwater pools at reactor sites for up to three years, then shipped away for reprocessing and final disposal. Although some reprocessing was done at a facility in West Valley, New York, before it closed in 1972, it did not occur as planned at other sites, and for the most part the reactor sites had to keep the waste in the pools.[27]

Yucca Mountain also involves geologic disposal but is technically different from WIPP. It is not an underground salt formation but a volcanic rock considered more suitable for containing the intense heat of high-level materials. But Yucca Mountain shares with its New Mexico cousin a protracted history of scientific-political and federal-state conflicts. In 1987, the federal government decided to concentrate exclusively on the Nevada site as the long-term storage facility to the exclusion of all other locations and technologies. The manner in which Congress mandated such an act—removing all other possible locations from consideration— enraged Nevada lawmakers. They have staunchly opposed the department ever since, hiring their own scientists to study the mountain and raising technical concerns.

The purpose of this book is neither to advocate nor oppose WIPP. Instead it aims to shed some light on the protracted process that gave birth to and nurtured the project through three decades' worth of successes and setbacks. The project's history raises numerous political, legal, technical, and emotional issues, all of which deserve to be substantively explored. With so much nuclear waste piling up around the United States, all sides involved appear to agree on at least one thing—additional WIPPs are more likely than not.

1

"A Way . . . to Make a Buck"

1971–76: CARLSBAD AND WIPP

It was one of the most anticipated garbage pickups in U.S. history. Shortly before 8 p.m. on March 25, 1999, a tractor-trailer hauling three steel containers each about the size of a Dumpster left the heavily guarded gates of the Department of Energy's Los Alamos National Laboratory in northern New Mexico's high-mountain terrain and began an eventful 260-mile journey.

As the truck rumbled away from the place where the atomic age had been conceived 54 years earlier, jubilant Los Alamos workers lined the road to greet it. "There's a lot of pride in this, a lot of pride," laboratory manager Dennis Rupp told a reporter. But when the truck reached the outskirts of Santa Fe, about 15 miles south, the primary onlookers became environmental activists who had fought for years against the prospect of ever witnessing such a sight. One tried to block the vehicle's path with his car before New Mexico state police interceded; others waved signs and beat Tibetan shaman drums in protest. "You're evil!" one woman yelled at the driver.[1] The truck continued south, a state police escort and television news crews in tow. Finally, at around 4 a.m., nearly 500 weary but excited bystanders cheered as the caravan pulled up to a cluster of stark white buildings rising out of the barren expanse of scrub oak, coyote trails, and mesquite upholstering the southern New Mexico desert. The truck had arrived to deposit its cargo at the world's first permanent deep underground burial site for nuclear materials: the Waste Isolation Pilot Plant (WIPP).

Nearly half a mile underground was an environment radically different from the quiet desert above. A maze of dim and sepulchral tunnels stretched

Figure 6: *The waste shipment arrives at WIPP (Department of Energy photo).*

seven miles in length, all connected to a series of rooms that had been carved out of the salt formation. Bright fluorescent lights illuminated the passageways as well as computers, seismographs, and other equipment. Workmen buzzed about in modified golf carts, leaving clouds of fine dust in their wake like ghostly apparitions. They unloaded the containers from the truck and put them through a security inspection, a radiological survey, and a series of other reviews. Satisfied that the materials were acceptable for burial, they packed them into an elevator and used a forklift to unload them inside a room marked with a warning sign between two lengths of chain: RADIOACTIVE.

Later that day, Energy Secretary Bill Richardson proclaimed his satisfaction with what had happened. "This shipment represents the beginning of a long-overdue promise to America to clean up our nation's Cold War legacy of nuclear waste, and to permanently isolate this waste from people and the environment," Richardson said. "And it is only the beginning."[2] Joe Skeen, a Republican congressman from southern New Mexico who had worked for decades to put waste at WIPP, was even more melodramatic. "God almighty," he exclaimed as his staffers celebrated with champagne in their Washington office, "why did it take so long?"[3]

The first move of radioactive materials to a deep geologic repository in the United States had taken almost 28 years. During that time, there had been enough scientific and environmental studies to fill a library, thousands of hours of political debates, hundreds of regulatory actions, several dozen lawsuits, and countless protests. Six different presidential administrations, more than 20 federal and state regulatory agencies, and scores of scientists and activists had been involved. It had been, by far, the most drawn-out and acrimonious debate in modern New Mexico political history. To many of those on both sides of the issue, the fact that the plant was finally being used for its intended purpose seemed almost unreal.

All the activity might never have occurred if a politician named Joe Gant had not read a newspaper article and sensed an opportunity. The November 1971 wire story in the *Albuquerque Journal* described Kansas's vehement opposition to radioactive waste experiments at Project Salt Vault near Lyons, mentioning Kansas attorney general Vern Miller's threat of legal action. "There's a small war brewing between the state of Kansas and the Atomic Energy Commission," it said. The article went on to note that Representative William Roy, D-Kansas, suggested that the United Nations be asked to explore other possible sites in uninhabited areas of the world while the commission was examining locations in Oklahoma, Texas, and Louisiana.[4]

Gant, a New Mexico Senate member from Carlsbad, was intrigued. He wondered why New Mexico was not on the list. He contacted his friend Harold Runnels, a resident of nearby Lovington, who was serving his first term in Congress. Runnels was a conservative Democrat who had spent 10 years in the state legislature before being elected to represent southern New Mexico's Second Congressional District. Journalists described him as "a good old boy." Affable and folksy, Runnels got by more on personality than intellect; *New Times* magazine later singled him out as one of "the 10 dumbest members of Congress." He would attract considerable notoriety in 1973 for announcing that, even though he was a member of the House Armed Services Committee, he had been purchasing documents from someone in the Pentagon because he could not obtain them through normal congressional channels.[5]

Runnels was receptive as he listened to Gant. Part of Gant's persuasive power lay in his background as a chemist; he could speak with some authority on the area's geologic properties. He also carried considerable local political clout. A soft-spoken, courtly, and gracious man, he had moved to Carlsbad in 1934 after graduating from the University of North Carolina. He became active in politics, serving on the Eddy County Commission before being elected to the Senate in 1969. "Joe was a real old southern gentleman—real polite, real quiet," recalled Eddie Lyon, a resident of the city since 1950 who spent more than two decades as executive director of Carlsbad's economic development agency. "We used to call him 'Whispering Joe.'"[6] Runnels encouraged Gant to talk to the federal government. "I feel it is my duty and obligation to handle requests either from individuals, cities, counties or states when asked to do so," Runnels said in 1972. "I recognize the fact that New Mexico, and particularly the 2nd District, needs to provide jobs for our people."[7]

At the time, Carlsbad had already developed a knack for self-promotion. Such a trait had evolved from its inception and flourished as a result of its unique natural assets, its changing economic history, and its independent-minded philosophy. Years before alarm bells began ringing in the nation's consciousness about the dangers of nuclear energy, Carlsbad's ability to aggressively market itself to the federal government would help Gant and other jobs-conscious leaders land an enterprise they considered perfectly tailored to their location—an underground storage site for radioactive materials.

Remote, Isolated, and Entrepreneurial

Geography was one of the city's chief selling points. Sitting in New Mexico's southeast corner along the Pecos River, Carlsbad is several hundred miles from Santa Fe and Albuquerque, the state's largest communities. The drive between those places and Carlsbad is a lonely multihour trek across windswept, sparsely populated plains. A few mountain ranges and towns pop up along the route—including Roswell, the city infamous for its storied association with UFOs—but the scenery largely is defined by cattle ranches, farms, oil pump jacks, and highway billboards. Culturally and demographically, Carlsbad also is quite distant from the places most commonly identified with New Mexico. The rural region it anchors is

Figure 7: New Mexico state senator Joe Gant (Department of Energy photo).

known as "Little Texas," the last section of the state settled by white "Anglos" rather than Spanish colonists. Little Texas has more in common with the traditional-values hamlets of the Lone Star State than with the Spanish-influenced regions of northern New Mexico. "Spiritually," one local author observed, "we are still the Bible Belt."[8]

Carlsbad looks like a typical middle-American small town, with a hodgepodge of postwar ranch homes, neon-sign motels, strip malls, and fast-food restaurants. Nevertheless, its history is straight out of American frontier West lore. It owes its existence to a businessman-turned-rancher from New York named Charles Bishop Eddy, a late-nineteenth-century entrepreneur. Restless and ambitious, Eddy was so driven to succeed that he never found time for marriage, but he had a recognized talent for persuasion. "Eddy could dream up something, begin talking about it, would soon begin to believe in it himself, was then irresistible and could convince any skeptic," recalled William A. Hawkins, his general counsel.[9]

Eddy started the community, which originally bore his name, once he realized that water diverted away from the Pecos River was essential to a successful cattle venture. He joined forces in 1887 with another promoter, newspaperman Charles W. Greene of St. Louis, and the legendary former sheriff of nearby Lincoln County, Pat Garrett, who had gunned down Billy the Kid to the north in Fort Sumner. The three men formed a company, and Eddy began seeking investors. Two years later, he found James John Hagerman, a Coloradoan who had made his fortune in the mining business. After getting Hagerman to invest in irrigation dams and canals, creating the infrastructure for settlement, Eddy began to set his sights on a railroad line. By 1891, the first train came westward from Pecos City, Texas, on 89 miles of fresh track. Eddy, the town, began to grow, bolstered by its reputation as a place where settlers from the East could feel safe. Unlike many other western communities of the era, it discouraged saloons and gambling halls and their by-products: prostitution, gunfights, and disorderly conduct.

It didn't take locals long to continue Charles Eddy's zeal for economic development. At the turn of the century, soaking in mineral water was regarded as a fashionable cure for various ailments. When Eddy residents learned their water had the same chemical composition as that of a famous European health resort in Karlsbad, Czechoslovakia, they voted in 1899 to change the town's name to Carlsbad, hoping it would lure more visitors. The gambit appeared to work—the city's population nearly doubled between 1900 and 1910, from 963 to 1,736. And while the spa fad declined, the city's fortunes did not. An adventurous cowboy, James Larkin White, discovered the first section of an awe-inspiring underground geologic formation a few miles southwest of town in 1901. The discovery eventually yielded room after room of huge and spectacular conical towers formed by centuries of mineralized groundwater. It became considered such an important national treasure that President Calvin Coolidge proclaimed it in 1923 "the Carlsbad Cave National Monument." Seven years later, Congress redesignated it as Carlsbad Caverns National Park.[10]

Other underground treasures soon became apparent. A government investigation in 1925 found that Carlsbad's soil possessed abundant quantities of potash, or potassium salts, an important ingredient in fertilizer for agriculture. With Germany placing an embargo on potash exports during World War I, American businesses were anxious to find other supplies. The Carlsbad area's first potash mine was incorporated east of the city in 1930 as the United States Potash Company. By the end of World War II, New Mexico's potash industry, centered in Carlsbad, had far outstripped all other domestic producers, furnishing 85 percent of the total national production.[11] By 1960, Carlsbad's population exceeded 25,000, tripling in just 20 years from 7,100. In addition to the thriving potash business, Carlsbad Caverns had become a dependable tourist draw, and farming and ranching remained dependable income sources.

But within a few years, the city's economic prosperity began to cloud as new discoveries of potash in Canada led prices to plummet from $50 a ton to as low as $11 a ton. In October 1967, U.S. Potash announced it would cease operations at the start of the year. City officials watched hundreds of residents leave; the 1970 population dwindled to 21,297. To offset the economic losses, the city began making an aggressive pitch for retirees, touting its year-round temperate climate and low cost of living. "I thought Southern California had the lock on promotion, but this gang down here is very promotion-minded," said Ned Cantwell, the former editor and publisher of the *Carlsbad Current-Argus*, the city's newspaper. "They've had to be. The mine closing was a terrible economic shock."[12]

Carlsbad's Interest in Waste Grows

Within a few weeks of reading the newspaper article about Kansas's objections to waste storage in 1971, Gant had gotten a number of Carlsbad residents interested in the idea, in much the same way that Charles Bishop Eddy had done a century earlier in attracting investors. One of the community leaders Gant turned to was Louis Whitlock, a Texas native who had lived in the city for three decades and was director of the Chamber of Commerce. While Whitlock shared Gant's civic fervor, he was brash, outspoken, and opinionated. A local newspaper reporter described Whitlock as "the only man who can strut while sitting down."[13]

Gant also worked with the mayor, Walter Gerrells. A native of the city, Gerrells ran a men's clothing store that had been started by his father. He stepped into public life in 1964 as the only non-union city council member elected that year. After six years on the council, he ran unopposed for mayor, establishing himself as a no-nonsense leader intent on maintaining a tight grip over his city. "He was very thorough, very conservative," Lyon recalled. "When he was mayor, he ran all of Carlsbad. You didn't work for the city until you checked with Walter."[14] Gerrells took a proactive approach to civic leadership. "I feel a town is as good as its people, and we shouldn't wait for others to do the things that need to be done," he told the *Current-Argus*.[15] Like Gant and Whitlock, the jobs and income that such a large new business could attract intrigued him. "We recognize the fact that we have an extractive industry as our major economy," he said in 1976. "We feel it's important that we diversify—that we have, over the long haul, other types of income."[16]

In many ways, Gant, Whitlock, and Gerrells fit the mold of political figures in the West who regarded politics as the handmaiden of economic development. Unlike the eastern and southern United States, western states and communities have all been forced to compete for the same assets, whether they have been private or federal. Consequently they have tended to regard each other as rivals without forming the set of internal alliances common to other regions. In Carlsbad's case, the city's geographic remoteness and differences from other areas complicated forging any such alliances. Unlike those other areas, the city was able to maintain a degree of economic independence, thanks to the potash mines. At the same time, Carlsbad maintained an influential federal presence. In addition to the Park Service, which ran Carlsbad Caverns, the Bureau of Land Management oversaw much of the acreage surrounding the city, while the Forest Service supervised nearby Lincoln National Forest.

"The dependence on the federal government has been the central reality of western politics," historian Richard White has observed. "The weaknesses of western political parties, the huge landholdings of the federal government within the region, and the long period during which the West was an economic colony of the East have all tended to diminish state power and increase federal power. . . . As a result, successful western politicians have been those able to secure federal favors for significant constituents or constituencies."[17]

Gant, Gerrells, and Whitlock assessed what they had to offer the government. Carlsbad was arid and without a prolific aquifer as a water

Figure 8: Carlsbad mayor Walter Gerrells (Department of Energy photo).

source. It was distant from any metropolis and major interstates. It lacked a major airport but was serviced by a good railroad system. It had hundreds of miles of deep passages in its potash mines, as well as several generations of miners familiar with the challenges of working underground. With the caverns, the city had successfully demonstrated its unique subterranean assets could be turned into a lucrative profit-making tourist venture.

Perhaps its most important asset, though, was its firsthand experience with nuclear energy. In the late 1950s, the Atomic Energy Commission (AEC) began seeking peaceable uses for the atom through a program called Plowshare, after the biblical citation from Isaiah 2:4, "And they shall beat their swords into plowshares." The idea was to determine if the energy from nuclear explosions could be channeled into such industrial, scientific, and civilian uses as excavation projects, tapping groundwater and minerals. Plowshare's inaugural mission involved an experiment known as "Project Gnome," situated in a salt formation 25 miles southeast of Carlsbad. The location made sense, being south of White Sands Missile Range, home of the Trinity Site explosion in 1945. The commission took an open approach to publicizing the experiment; no less a figure than Edward Teller—father of the hydrogen bomb—assured citizens that the test would cause no harm. City officials began thinking of the potential economic benefits. A 1959 *Current-Argus* headline conveyed hope: "Atom Bomb May Be Boon for Carlsbad: Could Bring About Further Industrial Expansion in Eddy."[18]

At the Gnome site, scientists dug 1,200 feet into the salt to create a spherical room 15 feet in diameter. On the morning of December 10, 1961, all of the neighboring potash mines were evacuated as a precaution. Then a five-kiloton bomb was exploded with about one-quarter of the power of the one dropped on Hiroshima 16 years earlier. The blast caused the soil to jump eight feet in the air and generated seismic waves measurable as far away as Japan. The explosion melted more than 2,000 tons of salt, creating a cavity larger than the base of the U.S. Capitol dome and taller than an eight-story building. Gamma rays turned the previously grayish walls inside the cavern several startling shades of deep blue, yellow, and black.

But the goal of using the heat from the explosion as an electric power source was dashed when scientists discovered what happened in the aftermath of the blast. Radioactive steam in the form of white vapor escaped from the mouth of the shaft, and the cavity did not seal itself. Although scientists did obtain some useful data on recovering radio-active products and studying seismic activity, Operation Plowshare did not go down as a success. To this day, the Environmental Protection Agency (EPA) continues to draw water samples at the Gnome site, finding radioactive elements such as tritium and cesium at levels exceeding drinking water standards. Officials have determined the contamination has consistently remained confined to the site and has not spread to any residential wells. The samples are taken next to a refrigerator-size cement marker placed above the explosion site commemorating the experiment.[19]

Carlsbad's Cold War–era experience with nuclear energy gave it an important frame of reference on which to evaluate government-sponsored nuclear projects. Since that experiment was neither highly disruptive nor detrimental to the community's fortunes, it became a significant factor in ushering in the eventual acceptance of nuclear waste storage. "Project Gnome paved the way and made us aware of nuclear energy," Whitlock said. "We were familiar with that. So [nuclear waste] seemed like kind of a natural thing."[20]

As a result of that project, Whitlock and others perceived the prospective benefits of storage as far exceeding the potential dangers. They trusted the government enough to take adequate precautions and above all else sought to keep any hazards in perspective. At that time, the mishap at Pennsylvania's Three Mile Island plant was seven years from occurring, and details about the federal government's activities in managing the weapons complex remained a public secret. Consequently, as they saw it, there was no known countervailing body of evidence to argue against what they were exploring. Taking waste, then, was rationalized as an acceptable risk. "When you say something is dangerous, it has to be judged compared to other things," Gerrells explained years later. "There are five farm-related deaths every day; if that isn't danger, I don't know what is. But by no means is that a reason to stop farming."[21]

The developments with Project Gnome and Carlsbad's subsequent willingness to consider nuclear waste storage have parallels with other cities hosting controversial materials. Experts have found that communities that already have risky facilities tend to have local cultures that are willing to take on even more risks.[22] In a general sense, the keys to political and public acceptance in those places hinge on the familiarity with the technology used, in addition to whether the benefits to the public are clear and whether the practitioners can be trusted. Two years before Carlsbad began looking at nuclear waste storage, in December 1969, the U.S. Army decided to transfer deadly VX and GB nerve gases from Okinawa to the Umatilla Army Depot in Hermiston, Oregon. The move outraged many Oregonians, but Hermiston residents were 95 percent in favor of it. The town's reasons for support were straightforward: munitions and toxic chemicals had been stored safely at Hermiston since 1941, so the hazard was not unknown and the record of dealing with it was good. The agency involved with the storage, the army, was regarded with respect. And townspeople clearly saw the economic value of keeping hazardous substances at the depot.[23]

Making the Pitch to Washington

Satisfied that they could make a case for their city, Gerrells and Gant led a delegation of Carlsbad leaders on a trip to Washington in December 1971. They met with Runnels and other members of New Mexico's congressional delegation, who offered more encouragement. "I can envision New Mexico becoming the waste disposal capital of the United States," Runnels said later.[24] But the legislators also urged the contingent to enlist the support of Carlsbad's residents, as well as Democratic governor Bruce King. "This should be gone over carefully with the governor and with the people at the local level," warned Claude E. Wood, a top aide to Republican senator Clinton Anderson, in a follow-up letter to Gant.[25]

Gant arranged for city, county, and state officials and local businesses to discuss the nuclear waste facility. But he and other local leaders made a conscious and important early decision not to involve the public. That decision established an influential precedent for how the planning process would unfold. "We had to be extremely careful at that time," he recalled in 1992. "We didn't know the particulars of what the AEC wanted to bury." As a result, he contended, "Meeting behind closed doors was a good move. . . . If we had run to the media to tell them we were talking to federal officials about getting WIPP here, well, I think they would have blown it sky-high."[26] Such an approach arguably altered the entire quarter century of events to follow.

Among the private meetings held was one in March 1972 at the office of King's administrative assistant, Toney Anaya. In addition to Anaya, the participants included representatives of U.S. Potash Company, the New Mexico Bureau of Mines and Mineral Resources, and the state Department of Development. U.S. Potash was a subsidiary of Continental American Royalty Company of Oak Brook, Illinois. Its president, William Muller, had written to AEC officials a month earlier, contending that the consensus among local leaders to locate such a facility in its abandoned mine "is not only acceptable but highly desirable." Muller, however, shared Gant's concerns about ensuring no word leaked out. "All are in agreement that the manner in which the possibility of such a project being approved is to be made known to the public deserves the utmost attention and coordination," he wrote. "Therefore, a method of coordinating public announcements between the political groups, the AEC and our company will need to be developed if we are to maximize public acceptance."[27]

Anaya, who would later be elected New Mexico attorney general and governor, remembered that the Carlsbad representatives' interest in improving their economy dominated the discussions. "The impression at the time was that this was a project that was going to create jobs," he said.[28] Gant, for one, had already turned his initial interest into income. One consulting firm, Nuclear Engineering Company of Walnut Creek, California, took him on as a part-time consultant, with a salary of $100 per day plus travel expenses.[29]

Following the 1972 meeting in his office, Anaya began to inform other state officials about what was under discussion. In a "confidential" memorandum, he noted that U.S. Potash's interest stemmed from "declining employment in the potash industry. . . . This would be a method by which to employ additional individuals. Also, as they candidly stated, it might be a way for them to make a buck." Nevertheless, Anaya declined to rule out the idea. His memo concluded: "I think it should be stressed that no endorsement nor discouragement has been provided by the governor's office. I think we need to know more about the proposal as it moves along, and specifically, we need to have your technical advice as to the desirability of the proposal."[30]

The desire for open-mindedness was in keeping with King's philosophy. The 48-year-old governor, elected in 1970, hailed from southern Santa Fe County, where his family maintained a cattle operation. King was influential in Democratic circles, having served as speaker of the state House of Representatives and state party chairman. He was the archetypal cowboy politician—"folksy, unschooled, shrewd and garrulous," according to a veteran Santa Fe journalist.[31] A subsequent fellow governor, Bill Clinton of Arkansas, would compare him to Will Rogers when campaigning for president in New Mexico in 1992. King was anything but polished; among his best-known malapropisms was his infamous warning that a controversy unrelated to nuclear waste storage could "open a box of Pandoras." But his lack of slickness camouflaged astute political instincts. He knew how to build alliances and avoid making enemies.

In that regard, King took a cautious and pragmatic view of balancing economics with environmental concerns. He admitted that he felt an affinity for rural areas looking to better themselves. In dealing with the environmental harm caused by strip mining in northern New Mexico, for example, King sympathized with mining companies that wanted to use the controversial technique to employ greater numbers of people. "We've got to deal with that problem with two ideas in mind," he told journalist and author Tony Hillerman in 1971. "If we let the land be ruined, it's usually ruined forever. But we can't afford to forget the people who need a chance to make a living. You've got to decide how high the return has to be in terms of payroll to make it worthwhile to accept damage to the land." He pointed out that his state was chronically among the nation's poorest. "We still have something here in New Mexico that most parts of the country have lost," he said. "It's just now beginning to pay off for us economically. We're beginning to cash in more and more on the vacation and outdoor recreation business, and we're beginning to attract some of the industries that don't pollute and that are getting tired of living with other people's pollution. So, just on the pure economic basis, we've got to strike a balance now—not just between the damage an industry does against the payroll it will produce, but also against what other business it's going to be keeping away in the future."[32]

The resulting absence of political opposition at the state and local government levels enabled the Carlsbad waste seekers to press ahead. Even an initial state Bureau of Mines study on the potential of southeastern New Mexico did not deter them. "The overall view of the area and some of the associated problems do not appear to cause it to warrant favorable consideration," bureau director Don H. Baker Jr. wrote in a December 1971 letter to Gant. In particular, bureau officials were concerned that any storage project would interfere with the future extraction of resources, such as oil, gas, water, and potash—an argument that many critics would repeat in the years to follow. Despite his misgivings about putting nuclear waste in a natural resource area, Baker, like Anaya, did not shut the door entirely, deferring to political leaders in Washington. "I would suggest that additional geologic and engineering investigations would probably confirm the negative nature of the area," Baker wrote. "However, if our congressional delegation and the AEC should feel that a more detailed evaluation is in order, we would make a proposal for financial assistance to accomplish this."[33]

Figure 9: Carlsbad's Louis Whitlock (left) talks with New Mexico governor Bruce King and King's wife, Alice, during a visit with former President Jimmy Carter (Department of Energy photo).

The immediate task of further study, however, would fall to another government agency, the U.S. Geological Survey. The survey had been asked to do an extensive evaluation of all the existing salt deposits around the continental United States; after the February meeting in Anaya's office, AEC officials asked the survey to pay special attention to southeastern New Mexico. After a few months, the survey found that the salt deposits in the area appeared thick and stable. Even more important, it discovered areas in which no mining or drilling was occurring. In addition, since oil drilling started later than in Kansas and was under the control of the federal government, they found there were better records about its historic development.[34]

As such data were emerging, a shift in the federal government's waste disposal policy also was developing. In the spring of 1972, AEC officials decided to change the objectives of their geologic program away from permanent disposal of high-level waste to a so-called bedded salt pilot plant, where limited quantities of such waste could be stored, and retrieved, in a salt mine. The idea was to obtain data that could only be had from a full-scale facility but that would meet the precondition of starting to build an expensive facility. As Energy Department officials would later acknowledge, the move toward a retrievable waste storage site was done with the idea that it could effectively eliminate the kinds of safety arguments that arose in Kansas and derailed Project Salt Vault. In May, the commission announced it would deal with high-level waste through a retrievable surface storage site while using the Bedded Salt Pilot Plant to show the safety of underground disposal.[35]

At the urging of Gant and the other Carlsbad backers, King extended a formal invitation to the commission in July to look at New Mexico for waste storage. By then, Continental American Royalty had dropped out of the picture; it was sold that same month to the electronics combine Teledyne.[36] But it hardly mattered—the federal government was interested. Having been stung by Kansas's opposition, though, the commission was unwilling to be proactive. "The state people were told we were not coming to New Mexico until invited by the governor," recalled Owen Gormley, an engineer in the agency's waste management and transportation division. "And we said that, if invited, we would be coming with geologists, not bulldozers."[37] Persuaded by such a hands-off pledge—as well as by Carlsbad's persistent lobbying—King was willing to give his tacit approval. "His attitude was, go take a look," Whitlock recalled. "Bruce was too smart to get caught with an outright commitment early on. He'd say, 'Some of my friends support it, and some oppose it, and I'm for my friends.'"[38]

The Public Learns the News

It was not until the following month—nine months after the initial contacts between Carlsbad and the federal government—that New

Mexico residents were told what was happening.

On August 14, 1972, Dr. Frank K. Pittman, the AEC's director for waste management and transportation, held a news conference to outline his agency's decision to examine the area for a potential nuclear waste storage site, a project he estimated would cost $25 million. The waste would come mainly by rail, he said, with some truck shipments. Pittman emphasized to reporters that Carlsbad's stable and self-sealing underground salt was what drew his agency, not its local leaders' enthusiasm. He also stressed that the agency did not want to "bulldoze anything" and would solicit state and public support. "Hopefully, early next year we'll make a decision on where the pilot repository will be located," he said. "If all goes through on schedule, we'll have a pilot facility ready by about 1979 or 1980."[39] State officials cited questions over transportation and the effects on the surrounding area but expressed satisfaction that their concerns would be listened to "very strongly." State Development Department director William Simms told reporters that New Mexico would not take a position until more facts were known—a stand King echoed in responding to letters from inquiring constituents. "While I think the decision must first be made at a local level in Carlsbad as to whether to proceed with the project, I wish to assure you that I will continue having my state agencies follow this project very closely," the governor said.[40]

King assigned responsibility for tracking the project to the Governor's Committee on Technical Excellence, a study group that had been formed in 1970 to try to tie New Mexico's defense industries and nuclear activities with private industry. In a March 1973 letter to the panel's chairman, King promised not to micromanage how the committee should operate. But the governor pointedly added, "As a general conclusion, I think the committee can operate under the principle that the state of New Mexico is one of the most logical locations for the national repository."[41]

The state government's deference, both to the federal government and the local interests of Carlsbad, opened the door to waste storage in New Mexico. But even if King had raised objections at the outset, he might not have been able to kill the idea. In the early 1970s, the federal government held considerable sway over what was perceived to be in the nation's best interests—an attitude with which New Mexico was already familiar, having involuntarily become the host of the atomic bomb's developers during World War II. "The environment that existed back in the days when this was being first proposed was that the federal government knows what's best, and states weren't permitted to be assertive," Anaya recalled. "When it came to something like nuclear waste disposal, national security issues, the feds had to do what they had to do, and the states had to get out of the way." At the time, he stressed, the state believed it was not in a position to argue. "Generally speaking, we didn't have many people, so we didn't have much political power," he said. "But we did have a lot of land that could be used for things like the disposing of waste. We were also a state then, as we are today, with a lot of federal installations. And so everyone said, 'Why not?'"[42]

Handed such an opportunity, the Carlsbad contingent went to work. To learn more about nuclear waste, its leaders attended seminars, workshops, and forums around the country. They began emphasizing to locals how such a project might someday prevent their sons and daughters from having to leave town to find work. "We do not want to see our children, our graduates from our school system, go to Dallas, Texas, and get a job, or Los Angeles or New York City," Gerrells later told a congressional committee.[43] The mayor oversaw the efforts of the storage supporters, although he parceled out the work. "Walter was very good at delegating," Lyon said. "He would tell you what he wanted and then sic you on it."[44] Publicly, Gerrells and others stressed that their hosting of the site hinged on whether it could be proved safe—although there was never a detailed discussion of what "safe" constituted. From the outset, Gerrells was confident safety would present no roadblock, tellingly citing Project Gnome as a precedent. "The radiation from this stuff is real small," he told the *Current-Argus.* "There's absolutely no comparison between the amount of radiation from this and the radiation we got from the Gnome project."[45]

Gerrells's efforts to shape public opinion received a considerable boost from the local media, as the *Current-Argus* maintained an open-minded editorial stance toward the idea of waste storage. Editor and publisher Cantwell helped defend Gerrells and others from criticism. Two days after the August 1972 news conference announcing the project, Cantwell wrote a front-page editorial responding to Kansas governor Robert Docking's comments that he was glad the AEC was focusing on Carlsbad. Docking had been quoted in an Associated Press story as branding the idea of waste storage as "unsafe" and as believing King would share his concerns. "While we appreciate the good governor's concern, we want to assure him that the people of New Mexico are capable of making their own decisions without the help of Kansas," Cantwell wrote. "In other words, mind your own business, Governor."[46]

The *Current-Argus* echoed city officials' views that it favored careful study of the idea. Stressing caution and safety appeared to be intended to avoid repeating the lessons of Lyons, Kansas, where safety concerns had undermined Operation Salt Vault's political legitimacy. Editorially, the paper derided the notion that the city supported storage at all costs but made little secret of the fact that it saw the most formidable obstacles ahead as emotional rather than technical. "Before these things begin happening," it warned in a July 1973 editorial, "we ask only that the people of Carlsbad continue to look only at the facts, thereby resisting the temptation to be influenced by scare tactics or wild speculation."[47]

The Scientific Search for a Site

As Carlsbad's proponents sought to shape public support, scientists began to intensify their scrutiny of outlying areas. By December 1972, four months after announcing the investigation into southeastern New Mexico's suitability, officials at Tennessee's Oak Ridge National Laboratory were already impressed with the area's potential, although they dismissed the idea of using abandoned potash mines for disposal. Of the salt beds in the Permian Basin—which encompasses Texas, New Mexico, Nebraska, and Kansas—the New Mexico area "appears to be most promising," according to an Oak Ridge report. "The principal factors leading to this conclusion are the large amount of geological data available because of the potash mining industry in the region; the presence of extensive and thick beds of high-quality salt located at appropriate depths; the possibility of disposing of excess salt in the existing potash mine excavations; the ideal surface conditions of low rainfall and low population density; and the fact that most of the area is currently in federal ownership." Yet the scientists echoed the earlier warnings of the state Bureau of Mines in citing "an appreciable potential for future petroleum exploration." They also found possible hydrological problems but added, "Preliminary information suggests that they can be avoided by careful site selection."[48]

After looking at and rejecting four locations, the scientists settled on an area known as Los Medanos, about 30 miles east of Carlsbad. The area was in the north-central section of the Delaware Basin, a geologic region of nearly 9,000 square miles stretching across western Texas, southern New Mexico, and Mexico. Los Medanos's distance from the city limits helped give the city's leaders an added measure of confidence. As they saw it, anyone from Carlsbad who would say "not in my backyard" to nuclear waste would have to point to a desolate prairie. In March 1974, field investigations at Los Medanos began with the drilling of two "core holes" at the northeast and southwest corners of a three-mile site.

Within two months, however, activities at the site were suspended. Atomic Energy Commission chairman Dixy Lee Ray refused to set aside the Los Medanos area or its protective "buffer zones" for the work to continue. Under law, waste repositories had to be on federal land, so before work could proceed, a portion of the property had to be transferred between the Bureau of Land Management and the commission. But such a move meant closing the area to oil exploration—and oil was a politically charged topic at the time. The Mideast oil embargo had led to widespread fuel shortages and long lines at gas stations, and any move to ban oil drilling would have been fraught with controversy.[49]

After Ray denied a request to withdraw public lands to prevent oil exploration on them, commission officials elected to explore general programs in geologic disposal rather than commit to building the Bedded Salt Pilot Plant. With the OPEC oil embargo under way, Ray did not want it to seem as if the government favored nuclear energy over oil and gas. That did not mean Ray opposed storing waste in the state—she had, in fact, already been recruited by Carlsbad's boosters. Early in their efforts, Whitlock and state representative Walker Bryan met with her at the commission's headquarters in Germantown, Maryland, outside Washington. Their decision to directly contact her would set the tone for Carlsbad's longtime lobbying strategy of bypassing lower- and middle-level managers in favor of those at the federal government's highest echelons. "It's been my experience that you go to the top to get decisions," Whitlock explained. The visit left Ray pleasantly surprised, and she spent the afternoon chatting and drinking wine with them. "She couldn't believe that two people would actually be coming and asking for [the project] when they were being run out of Kansas," he said.[50]

Subsequent outside events, however, would move the work in New Mexico off the back burner and nudge it closer to reality. In trying to deal with the oil crisis, new president Gerald R. Ford saw a need to reshape the government's energy policy. In October 1974 he signed the Energy Reorganization Act, abolishing the AEC and establishing two new bureaucratic entities: the Energy Research and Development Administration (ERDA) and the Nuclear Regulatory Commission (NRC). The NRC was handed the responsibility of licensing and regulating high-level nuclear waste repositories and overseeing the commercial nuclear power industry, while ERDA was left to carry on AEC's other activities. Although the new agency's waste management officials were for the most part the same as the commission's, they brought with them the potential for a change of direction. As employees of the new agency, they could look at nuclear waste disposal in a new and different way.[51]

One of the earliest actions affecting New Mexico was to move the management of the waste storage project from Washington to the control of the ERDA's Albuquerque Operations Office. Sandia National Laboratories in Albuquerque, meanwhile, was given technical responsibility for developing the repository. Sandia had experience in setting up nuclear-related test facilities in other remote parts of the West, including California's Salton Sea and Tonopah, Nevada. Oak Ridge officials were unhappy at having lost a major project and expressed concerns that Sandia would be "reinventing the wheel" by starting a new investigation. Albuquerque officials stressed they would draw on Oak Ridge's expertise and research.[52]

Officials at the new energy agency also were eager to take a second look at the overall waste storage concept, particularly the retrievable storage facility idea that the AEC had proposed. It had come under criticism; the EPA feared that institutional inertia would lead to problems and gave it the agency's lowest-possible rating. Faced with such a situation,

ERDA officials decided to scrap the idea of a retrievable facility and focused exclusively on the New Mexico site in salt, the geologic medium
that the National Academy of Sciences had found so promising two decades earlier. By 1976, that project would bear a new name: the Waste
Isolation Pilot Plant.[53]

In Carlsbad, local boosters were pleased with themselves for putting
their idea of "making a buck" in motion. They expressed satisfaction
that local opposition had not derailed the federal government's efforts
to investigate nuclear waste storage. "All the community leaders, from
the very inception, were in lockstep for this project," recalled Whitlock.
"When you have that kind of cooperation, you're going to move
forward."[54] But residents and officials in other areas of New Mexico would
not prove so cooperative, and before long their interests would come
into direct conflict with those in Carlsbad.

2

"This Is Not a Predictable Situation"

1975–78: OPPOSITION EMERGES IN NEW MEXICO

Sandia National Laboratories scientists did not expect to be startled in June 1975 when they made their first foray into the field to obtain data for the Waste Isolation Pilot Plant (WIPP). They picked a site east of Carlsbad known as ERDA-6, in the northwest corner of an area that Oak Ridge National Laboratories already had mapped out. But the Sandians soon learned that the geology of the area was at odds with what they had believed.

Data from drilled boreholes showed the underground salt beds dipped steeply at angles up to 75 degrees instead of being level and gently inclined at the angle of one or two degrees in nearby beds. The deformities were a drawback for federal Energy Research and Development Administration (ERDA) officials, who were contemplating a nuclear waste repository on two levels, with one above the other. The main level would be used for the transuranic waste accumulating at weapons production facilities in Idaho, Colorado, and elsewhere; the second level was to be studied for storing high-level spent fuel from commercial power plants. A memo was sent to agency headquarters that Sandia "has made the technical decision that it is not feasible to continue drilling operations in light of the dangerous potential existing in the hole."[1]

The scientists encountered another disturbing development. When the drillers reached 2,710 feet at ERDA-6, they hit a huge pocket of underground brine—salt-saturated water—that gushed forth. The brine pocket was found in fractures of anhydrite, an evaporative rock often found in layers alternating with salt. Anhydrite, unlike salt, is brittle: when it deforms, it can fracture, causing brine to accumulate. Dissolved in the brine were hydrogen sulfide, a highly toxic gas, and methane, which is explosive. One of the Sandia workers was almost killed when he inhaled the gas; he received emergency oxygen and was rescued at the scene. The brine was regarded as a serious hazard. "If anything went wrong, it would hit the headlines," William Armstrong, an engineer on the project, told a reporter a few months later. "If we hit a gas brine pocket and killed a few people, hell, it would be spread all over the front pages."[2]

Although that never occurred, WIPP's unfolding technical situation would, in fact, be spread over the front pages. News about the project's developments would open the door for continued conflict between an evolving environmental movement and a newly reorganized federal bureaucracy unprepared to begin a new project under a public microscope. As citizen confidence in government and the nuclear industry began to sour during the 1970s, increasing numbers of activists objected to the official scientific line. Their challenges were grounded in deep skepticism on two parallel fronts: the technical merits of the site and the broader issue of nuclear usage. Because the technical questions did not always yield immediate or clear-cut scientific answers—and because proponents of the project did not always distinguish between serious technical skeptics and antinuclear activists in dealing with issues that each raised—a climate of scientific doubt and political divisiveness arose. In such a hostile and polarized atmosphere, consensus would prove impossible.

Before Sandia ventured into the field, a few experts had forecast that southeastern New Mexico might prove geologically problematic. Two years earlier, New Mexico Institute of Mining and Technology geochemistry professor Gale Billings warned that one of the government scientists' basic assumptions—that salt layers can seal cracks and are waterproof—was not always the case in the area under study. Around Carlsbad, Billings said, the salt already had caused water-filled cavities near the surface.[3] Even the U.S. Geological Survey, whose research helped propel early efforts in studying southeastern New Mexico, would subsequently cite concerns about brine in burying nuclear waste. "If relatively small amounts of brine can cause . . . possible movement of waste during a relatively short time, special efforts will surely be necessary to insure retrievability from a salt repository for periods as short as 10 to 25 years," the survey said in a 1978 study. "The question of whether the workings of a mine in salt can be predicted to stay dry will have to be faced." It recommended that other forms of rock be examined for burying high-level wastes.[4]

At the WIPP site, officials would eventually learn that brine flow was a common occurrence for oil and gas drillers in the Castile formation, the underground rock formation underlying the salt beds at the repository level. Of the 60 oil or natural gas wells drilled in the vicinity of WIPP, 10 of them hit brine in the subterranean zone.[5]

Sandia's Weart Takes Control

When the ERDA-6 location was deemed unsuitable, Sandia was faced with breaking the news to Carlsbad. The task fell to Wendell Weart, the laboratory's 43-year-old WIPP project manager. Weart had become involved in drilling the original test holes a year earlier and had been named to Governor Bruce King's technical advisory commission to review Oak Ridge's work. Weart was an Iowa native who received his Ph.D. in geophysics from the University of Wisconsin in 1961. That same year, he worked on Project Gnome, familiarizing him with southeastern New Mexico's geology. As the man in charge of underground motion measurements, he was among the first to venture inside the massive Gnome cavity after the subterranean explosion.[6]

A patient and low-key scientist, Weart had an authoritative manner and a knack for putting people at ease. He could explain complex issues in layman's language, inspiring confidence in his judgments, and his small-town Midwest background enabled him to relate to rural New Mexico residents. His skill in outlining the technical problems to anxious Carlsbad boosters greatly impressed his colleagues. "I don't know of anyone in the laboratories that could have handled it as well as Wendell—[it brought] tears in my damned eyes what this guy [was] doing," recalled fellow Sandia scientist George Griswold.[7]

Weart became—and remained for two decades—the person most closely associated with the scientific side of WIPP. He spoke to schools and civic groups across New Mexico and west Texas, appeared at public and technical forums, served as the principal contact for news reporters, and maintained correspondence with researchers in other nations—all while remaining the federal government's liaison with the scientists on-site. In reflecting on the project's early years, he noted that "my job was as much public relations, and public information, in those days as it was trying to direct the scientific studies."[8] In retrospect, he believed such a task could have been more successful if he and others had couched their early assessments in more ambiguous terms. "The biggest mistake we made was, before it became an issue, telling people that one of the things we liked about salt was that it's dry," he said in 1999. "The minute the first beads of moisture appeared, we had to start telling [outsiders], 'It's still a dry rock in mining terminology; it's just not bone-dry.'"[9] The public could, and did, interpret such statements as a scientific retreat.

Weart provided institutional continuity on the scientific side that would prove lacking on the political end. He did not dictate policy, although his memory and knowledge would prove beneficial to the succession of officials who did. His unflappable manner would be a reassurance to WIPP's proponents. "Wendell is the type of person who can dignify any question that's asked of him," said Eddie Lyon, who was promoting the project as head of Carlsbad's economic development agency. The chairman of the National Academy of Sciences' WIPP panel from 1984 to 1997, Charles Fairhurst, later described Weart as "an exceptional scientist who has brought international respect and admiration to Sandia for its role in WIPP."[10] By the time he retired from Sandia in 2000, Weart had acquired an affectionate nickname: "The Sultan of Salt."

In hindsight, however, Weart acknowledged that project officials were unaware of the degree of external scrutiny that WIPP would receive. They were still operating under the assumption that a safe site would be easy to find and that the public would readily accept their arguments. "When we started this project, we expected it would take maybe half a dozen years, and then we would turn it over to the operator," he said in 1999. "It turned out to be a much more difficult process to provide the necessary degree of assurance. We were a little naive in thinking what it would take to do a job of this sort."[11]

Changing the Site Criteria

Once ERDA-6 had been discarded as a potential site in 1975, Weart became one of a handful of scientists involved in the decision to relocate to a nearby area. Despite the problems, those scientists were convinced the region remained more promising than sites in other states. The end result was a shift six miles to the southwest, to a place both Sandia and the U.S. Geological Survey regarded as more geologically stable. Investigations began in early 1976, and by June, Weart told a reporter, "We are pretty confident we have a predictable and normal geologic situation."[12] The new area, however, was considerably closer to oil and gas boreholes. As a result, officials chose to relax the guidelines on how close those wells had to be from two miles to one mile so the number of potential sites could be widened. The scientists were confident that one mile remained a sufficiently comfortable distance. But as *Science* magazine reporter and author Luther Carter observed, "The circumstances invited suspicion that criteria were being picked to accommodate the available sites rather than vice versa."[13]

The federal government's desire to formally select an official home for WIPP did, in fact, nudge along the scientists' decision. In June 1976, Sandia's Melvin Merritt told a reporter, "There is pressure from Washington to tell them in the next month or so. The pressure is to provide a status report by the second week of July" in time for a national waste disposal conference.[14] Sandia quickly complied, although Weart denied politics outweighed scientific considerations. "It was really selected on the basis of this geological setting being superior to the other sites," he said. "As you went north, the salt beds became thinner, closer to the surface and more susceptive to groundwater." In a June 21 memo, Weart conveyed the news that "on the basis of the geological and geophysical work completed to this point, the area currently under investigation for the Waste Isolation Pilot Plant meets existing site selection criteria and is acceptable for the next stage of site development." He asked that action be taken "as soon as possible" to protect the site from oil and gas drilling. Six months later, in December, the energy administration filed an application with the Bureau of Land Management to "withdraw" 17,200 acres of land in Eddy County, segregating them from public entry.[15]

Sandia was given the task of issuing what would become the first public document on WIPP: a draft environmental impact statement. The document spelled out in general terms what was proposed for the site, after laboratory officials spent months researching its climate, geology, and natural resources. It called for accepting and storing nuclear materials "in limited but realistic quantities and at realistic rates of receipt." Both transuranic waste from nuclear weapons plants and commercial high-level wastes would be sent by rail or truck, with the transuranic site situated 2,100 feet underground and the high-level area another 500 feet below.

In addition to providing the first specifics of storing waste at WIPP, the document revealed the prevailing governmental impatience toward finding a home for the waste. As was required under the new federal National Environmental Policy Act (NEPA), the impact statement examined alternatives to underground storage in New Mexico, including taking no action at all. But it dismissed those options. "There are a number of alternatives that involve no action at this site; some of them are possible but would cost time and money with no assurance of a better program," it said. "Delay is a viable alternative, but one must remember that delay in general does not change risks, it only permits them to be better known. It usually costs money." It also acknowledged it was impossible to claim that WIPP "is the optimum design in the optimum location. There has not been the exhaustive investigation of alternatives that would be needed to make this claim; even if there had been, such an assertion would be inherently unprovable and hence, unwise."[16]

Officials at ERDA rejected the draft for what amounted to political reasons. The problem lay with the description of the plant's mission as an experimental facility, which until then had never been fully outlined. An ERDA representative explained to Sandia that "the primary audience for [an environmental study] was Congress, and we just couldn't go to them and say we had a half-billion dollar experiment to propose—even if this presentation would go over a lot better in [New Mexico]. We should rather take the positive attitude that the ERDA intended to build a repository" after an initial, limited period in which the waste could be easily retrieved. One official in Washington complained that "some of the descriptions of the project [in the study] leave the impression that we know very little about safely isolating non-heat producing nuclear waste."[17] Such assertions, when made public a year later, would come to be regarded as prime examples of how political decisions could conflict with, and override, scientific views.

The problems in issuing the environmental study reflected how WIPP's mission was subject to constant change during the late 1970s. Sometimes it was considered for storage of transuranic materials; at others, it was regarded as a burial ground for commercial high-level spent fuel. At the time, the fluidity reflected federal officials' desire to build in the flexibility to allow for a future expansion. But they did not articulate their strategy to the public, creating confusion and fueling controversy. As Darleane Hoffman, an official at Los Alamos Scientific Laboratory, wrote in a 1978 presentation to a Senate subcommittee: "The purpose of the WIPP site was changed in midstream, leading to assertions that the government did not know what it wanted, or still worse, was not telling all of its intentions. And people asked, if the purpose of the site could be so easily changed, what was to prevent further changes?"[18]

The DOE and Rise of Public Opposition

The successor to ERDA would not prove any more successful in articulating a vision that could win wide acceptance. In October 1977, the Department of Energy (DOE) became the federal government's twelfth cabinet-level agency, demonstrating Congress's and President Jimmy Carter's increased concerns over energy. The new department encompassed ERDA as well as the Federal Energy Administration, the Federal Power Commission, and assorted energy-related components of the Interior and Commerce departments. The organization act called for the department to establish control over all federally owned and managed wastes, as well as all commercial wastes not stored at nuclear power plants. It also commanded the department to set up programs to treat, manage, store, and dispose of the wastes. Both civilian and defense waste management functions were within the new department's Office of Energy Technology, although the agency's new secretary, James R. Schlesinger, promised to move waste management within the Office of Environment. Schlesinger, a former secretary of defense and head of the Atomic Energy Commission (AEC), believed there would be "a greater competence and, perhaps, a greater pursuance of that program if we make the change." But nothing came of the idea during Schlesinger's term.[19]

The Department of Energy's creation coincided with the rise of public activism on WIPP. As the environmental impact statement and other documents began to allow for outside opinions, and as on-the-ground site progress fueled the news media's curiosity, the project's critics became an entrenched and unyielding adversary to a bureaucracy that never before had to consider the public in its decision-making processes. After the Vietnam War and the Watergate political scandal, public confidence in government had taken a battering. One of the initiatives left to deal with the aftermath was WIPP.

The relentlessly skeptical attitude of the project's critics—with their "prove it before you can proceed" attitude—clashed with the scientific and governmental optimism that disposing of nuclear waste could be dealt with just as successfully as building nuclear weapons had been. In what was to become a long-running argument, the critics justified their stance on the grounds that a deep geologic repository for nuclear waste had never been built before. The Energy Department and Sandia said in response that did not mean they were flying blind, since they had accumulated useful scientific experience. But the absence of specific proof that a project such as WIPP could be built meant that establishing trust was particularly crucial—and trust was in short supply.

New Mexicans had raised questions almost from the project's inception. But two obstacles hampered them: a lack of available information and a suitable forum in which to express their misgivings. The bureaucratic culture of federal energy agencies was not geared toward proactive public involvement; official hearings were nonexistent. As a result, much of the early debate was over the general idea of waste storage near Carlsbad rather than on the technical merits of a particular location. In early 1973, a reporter for the *National Observer*, William Lanouette, concluded in an article about the newly proposed project that most of the city's residents "are skeptical and taking 'a wait-and-see' attitude."[20] The *Current-Argus* dismissed Lanouette's attempt to gauge opinion. "The writer does not bother to point out that the official response by Carlsbad to the AEC, a position endorsed by this newspaper, is that Carlsbad is fully in support of a

PILOT PROGRAM, a study to find out if burying atomic waste in nearby salt beds is safe," it said in an editorial. The newspaper criticized Lanouette for "making Carlsbad people out to be hicks."[21]

But some Carlsbad residents were indeed skeptical. Among them was Roxanne Kartchner, who was regularly quoted in articles in the national media in the late 1970s. In part, the attention came because she was a young housewife with roots in the area, not a stereotypical activist. Kartchner recounted to *Playboy* magazine in 1979 how Mayor Walter Gerrells told a friend of hers that WIPP "was none of her business, none of the town's business. The mayor was angry and upset about it. That made me curious. Anything that goes on in this town and this country is my business. I wondered why WIPP was such a tight, closed little arrangement."[22] Kartchner, who served as spokeswoman for a group called the Carlsbad Nuclear Waste Forum, helped collect signatures from other locals opposed to the project but discovered most of them believed a challenge was fruitless. She lamented that "the power structure has stuck together like glue on this issue."[23]

By this time, Gerrells and other boosters had formed their own group, Carlsbad Citizens for Energy Development. Bank president Mike Levenson led the organization, which boasted about 75 members, many from the business community. It held educational meetings around the city and in neighboring communities to answer questions about the project and tout its economic benefits. The group echoed the *Current-Argus* in portraying itself as open-minded toward storage but simultaneously concerned with safety. "We're in favor of WIPP whether it is in Carlsbad, Nevada or on the White House lawn—wherever scientists can prove it belongs," Levenson told a reporter.[24]

The group succeeded in winning the support of a coalition of lawmakers and businessmen in Lea County, east of the WIPP site, in early 1979. Endorsements from other cities and counties would follow, as Carlsbad's boosters touted the idea that the site would be used to help solve a national problem and that it was the area's patriotic duty to help. "We have realized that this project is of much more magnitude than we originally thought—that it is not only a local and a state concern, but it is important to our nation in its continued energy development," Gerrells told a congressional committee in 1979. "We realize that it is part of the nuclear energy cycle."[25] Such arguments helped persuade leaders in neighboring communities and, the group insisted, increasing numbers of Carlsbad residents.[26]

An Albuquerque Group's Growing Influence

If such lobbying helped mute criticism in southeastern New Mexico, it had little effect in the more geographically distant and politically liberal cities to the north, Santa Fe and Albuquerque. Unlike Carlsbad and southeastern New Mexico, where debates over WIPP frequently centered on jobs and economic impact, Santa Fe and Albuquerque residents perceived themselves as having to share the burdens and hazards of waste storage without receiving any of the corresponding benefits. The debate became a regional New Mexico "north-versus-south" conflict, in which two disparate groups fought to make their position known to the rest of the state and to policy makers in Washington.

Leading the criticism was the Southwest Research and Information Center, a nonprofit public interest organization established in 1971 during a period that saw environmental activism spreading across the nation. The center was the brainchild of University of New Mexico architecture professor Peter Montague and his wife, Katherine, who started the group in Albuquerque after working for consumer activist Ralph Nader. Intelligent and idealistic, the Montagues intended Southwest Research not as a traditional proactive organizing entity but as a technical assistance organization for communities with environmental problems. Raising money from private and public foundations as well as through individual donations, they tackled such issues as Indian water rights and sewage treatment before learning of Carlsbad's waste storage plans. They began demanding information. "We hate to see people being snowed by the AEC so they have the impression there are no problems, no dangers," Montague told the *National Observer's* Lanouette in 1973.[27]

As Montague learned more, his academic background and environmental credentials established him as leader of the opposition. His stance, in turn, caused his group to shift into an organizing mode. His initial opposition was grounded in a broad mistrust of nuclear power. "It dawned on us that if we could make waste disposal the focus of attention, that so long as we could keep waste out of the ground it could keep nuclear power plants from opening," he said in 1999. That belief, obviously, left him uninterested in achieving consensus unless it was on his terms. Unlike the Carlsbad boosters, he remained suspicious of Sandia's Weart, his occasional debating partner on radio talk shows and public forums and the frequent counterpoint to his assertions in the news media. "I think Wendell believed that the best thing to do was to get this waste in the ground," Montague said. "He was willing to make any argument necessary to achieve that goal."[28]

For Montague and other environmentalists, it became a bedrock belief that the federal government was intent on opening WIPP at all costs. They would conclude that as a result, project officials were unworthy of trust, no matter what was said or done. As studies have noted, public opposition to nuclear waste sites is considered far more likely when the implementing agency is viewed as lacking credibility.[29] A perceived lack of attention to the public information process compounded the government's problems. The situation allowed Southwest Research to advance an argument it would repeat for decades: that the site's technical deficiencies were compounded by equally flawed methods of involving the public. "No goals have ever been established for this project, as far as we know," Montague told a congressional committee in 1978. "If someone, somewhere has defined goals for WIPP, it certainly was not done by any public or open process."[30]

To bolster arguments against WIPP's technical merits, Montague cited the Sandia decision to relax the criteria on oil and gas boreholes, accusing the laboratory of bowing to expediency. He also began questioning the safety of transporting wastes. Montague sent an assistant out to photograph highways around the state that were likely candidates for shipping routes. (The WIPP environmental impact statement did not include any formal candidates for the routes.) What he found were roads with narrow shoulders, cracked and deteriorating pavement, and other problems.[31] At the same time, the group raised the prospect of high-level waste being stored permanently at WIPP, something Weart maintained would be done only on an experimental basis. By thrusting such issues before the public and news media, Montague helped to create the climate of scientific doubt that would make it difficult for the government to win public support. In September 1978, a Zia Research/KOAT-TV opinion poll showed respondents opposed WIPP by a 2-to-1 margin.[32]

A Geologist Joins the Fray

Montague's efforts received a lift from a scientist who lent credibility to his organization's arguments. Roger Anderson was a genial antiwar intellectual known for holding discussion groups about Vietnam at his Albuquerque home. But he was also a University of New Mexico geology professor and contractor for Sandia on WIPP whose specialty was studying the evolution of climate patterns. He had worked extensively around Carlsbad; a decade before the brine pocket accident at ERDA-6, he and a student had done some drilling not far away and also hit brine and hydrogen sulfide gas. He was attempting to count and measure each layer in the Castile formation, year by year, from bottom to top, to assemble what he described as the longest annual record of climate change in the world.[33]

In January 1978, Anderson submitted a report to Sandia that concluded the proposed WIPP site was in one of the best spots of the Delaware Basin, the large underground formation of western Texas, southern New Mexico, and northern Mexico. But Anderson warned that the entire basin suffered from problems that could plague the project. In the lower Salado salt formation, the underground area where Sandia's drillers had struck brine at ERDA-6, Anderson found evidence of groundwater flows dissolving the salt. He calculated that over the last 4 million years, water had removed 73 percent of the lower Salado's salt. If the same rate of dissolution continued, he estimated that groundwater could flow into the waste in 1 million years—and it was likely the process would speed up as time passed.[34]

Anderson's report also described "collapse features" directly beneath parts of the site. Such features were places where flowing groundwater dissolved salt. The resulting space in the earth had been filled by salt moving downward, often leaving a sinkhole or depression on the surface as well as humped mountain-shaped structures in the salt known as anticlines. He predicted that if such collapse features were present in the salt beneath the site, it would present "a short-term geologic hazard" because of the potential for pressurized brine and hydrogen sulfide gas. He also noted that broken rocks associated with such collapse features could form a conduit to transmit brine from the area and potentially release waste to the environment. In concluding his report, he warned that detecting such small collapse features "is not possible in the present state of the art. . . . The potential for the formation of collapse structures in the site area constitutes an unavoidable geologic hazard that is inherent in any site selected in the [Delaware] Basin."[35]

Anderson continued to refine and advance his arguments, enlisting the support of other geologists. They argued that more study was needed to determine why there was so much extensive salt dissolution in the area. By 1980, four scientists who also had been involved in WIPP research joined him in urging Governor Bruce King to hold a hearing on the problems. They told the state Radioactive Waste Consultation Task Force in April 1981 that WIPP had not yet been shown to be geologically acceptable, invoking what they saw as the uncomfortably high degree of scientific uncertainty. One of the four scientists, New Mexico Institute of Mining and Technology hydrology professor Lynn Gelhar, argued that the public was looking for a site with predictable geologic conditions. "To me, with the current state of knowledge," Gelhar said, "this is not a predictable situation."[36]

Energy Department officials strongly disagreed. They regarded the salt dissolution and brine issues as worthy of study, but not reasons to abandon the site. Joseph McGough, WIPP's project manager, told the state panel that brine reservoirs did not compromise the project. Without further exploration underground, he added, the agency was "at a point of diminishing returns with regard to the informational value of additional geotechnical data that can be obtained."[37] He and others pointed to the National Academy of Sciences, which had joined the scientific debate in March 1978 at the Energy Department's request. The academy panel of independent technical experts agreed that continued investigation was warranted. In a February 1981 memo to the members of the panel, Oklahoma mining engineer D'Arcy Shock concurred that Anderson's criticisms should not deter WIPP. "His objectives may appeal to many who, for one reason or another, hope to stop the project or who honestly feel that more caution is needed," Shock wrote. "However, it would seem prudent for the panel to review Anderson's reasons sufficiently to give its opinion of the importance of his objections. I see good reason to continue to encourage investigation of the existence and mechanisms of deep solution under salt strata, so that the apprehensions of disaster can be put into proper perspective."[38]

Indeed, the brine issue would be examined for years. From time to time, it would become the focus of public anxiety, most notably in November 1981 when a borehole one mile north of WIPP's central shaft was deepened. Drillers encountered a pressurized brine reservoir much larger than the one at ERDA-6, yielding more than 1 million gallons of brine, prompting critics to dub it "Lake WIPP." After that, the Energy Department agreed to state officials' request to rotate the room plan 180 degrees to place the repository south of the central shaft rather than to the north. The event gave critics more ammunition: one group known as Citizens Opposed to Nuclear Dumping in New Mexico clamored for "a full disclosure" of facts about brine problems.[39] Two years later, Weart summarized Sandia's evidence on brine reservoirs. His report concluded that their presence was unlikely, although it conceded it was "virtually impossible" to prove it. "Brine reservoirs are not likely to occur at the site now or in the near geologic future," the report said. "If they should occur, they will not interact with WIPP except through human intrusion and the consequences of this (unlikely) occurrence are not unacceptable."[40]

Anderson was not easily mollified. As he continued his research, he found another suitable candidate for a storage site where he did not believe dissolution and brine flow posed as much of a problem. But the site was in Texas, a more politically powerful state than New Mexico. Larger states, he and other critics lamented, had bigger congressional delegations and more influence with presidential administrations. He was dissatisfied with the National Academy of Sciences' handling of his concerns, believing it to be motivated to tamp down political opposition. Some scientists familiar with WIPP and sympathetic to Anderson suspected that his work was regarded as little more than speculation at the Energy Department. Among the geologists at Sandia, by contrast, Anderson was respected, even among those who disagreed with him.

Sandia terminated Anderson's consulting contract in 1981. By then, Anderson had become convinced that politics had superseded science in developing WIPP. "I've seen two distinctly different kinds of science being done," he said in 1999. "One is the pretty standard developing of a traditional hypothesis and trying to destroy it. That hypothesis and trying to knock it down still goes on, but most science is goal oriented, project oriented. . . . For the scientific part [at WIPP], it was a mission-oriented, military-type project where the goal was to complete the repository. The whole idea of doing the science first was backwards."[41]

The Failed Effort to Organize Opposition

Despite advocates such as Anderson, Southwest Research members still thought that harnessing a broad base of opposition was beyond their mission. To address the problem, however, Montague began hiring staffers to spread his message to the public. One was Jeff Nathanson, who had worked for the University of California Public Interest Research Group in Berkeley, the hotbed of 1960s antiwar activism. Another was Louis Colombo, who was doing graduate work in urban planning. The two began meeting with Montague and another Southwest Research member, University of New Mexico geophysicist Charles Hyder, in the fall of 1977 to develop a work plan. However, political inexperience, a lack of money, and other problems would hamper their efforts.

At the outset, the organizers were confident. The *Albuquerque Journal,* New Mexico's largest newspaper, had named the project the most important issue in the state. They developed a strategy of "coalition organizing" modeled after antinuclear campaigns in other states, including California and Oregon. The idea was to start in early 1978 to identify the attitudes of New Mexicans toward WIPP and develop tactics and a media campaign. The information would be coordinated with a media consulting and technical assistance organization. The result, they hoped, would be a group of anti-WIPP organizations that could form a broader, more politically powerful statewide coalition. To build such a movement, Colombo and Nathanson hoped to draw from low- and middle-income residents, as well as church, civic, and senior citizen organizations, across eastern and southern New Mexico. By starting in those mostly rural regions, Colombo said, they hoped "to give the anti-WIPP struggle a local, New Mexico character," which meant de-emphasizing urban, antinuclear groups. That approach, however, effectively circumvented the antinuclear and antiwaste work of two other groups: Citizens

Against Nuclear Threats and the Roadrunner Alliance.[42]

Other problems arose among the groups. As Nathanson and Colombo discovered, the new anti-WIPP organizations consisted mainly of "younger, liberal, college-educated people who were relatively unfamiliar with organizing or campaign work. This was the first sustained political activity for many of these people." In addition, some of the groups saw taking an anti-WIPP position as largely a symbolic action, one that required no active involvement. To varying degrees, all of the groups depended on Southwest Research for organizational support. But because of the center's tax status as a nonprofit organization, it was legally prohibited from investing a significant amount of time in political campaigns. The restrictions became a severe financial handicap and prevented the organizers from getting their network in place in time for the June 1978 primary elections in New Mexico.[43]

By the summer, Colombo had helped form a statewide coalition organization based in Albuquerque, known as Citizens for Alternatives to Radioactive Dumping (CARD). The founding of the coalition came several months after the researcher hired to examine New Mexicans' attitudes toward nuclear waste disposal in their state, Paul Fine of Oakland, California, issued his findings. Fine said his research found that many residents he interviewed believed the federal government would do whatever it wanted, regardless of the public's wishes. "This came out in an often-repeated phrase: 'They will shove it down our throats,'" he said.[44]

Hoping to capitalize on such attitudes, Southwest Research joined four Carlsbad groups in October in holding the first anti-WIPP rally in Carlsbad, which drew more than 200 people. That same month, it began running television advertisements to boost their organizing efforts. The advertisements depicted a nuclear waste truck, loaded with 55-gallon drums, traveling along New Mexico highways and streets. Images of the truck were crosscut with a group of children gathering tools and walking to tend a garden. The two themes converged when a hoe dropped out of a child's wagon directly into the truck's path. No crash was depicted, but the ad strikingly conveyed the critics' sense that unexpected events could occur and that nuclear waste transportation was risky. But the ads did not urge New Mexicans to talk with political candidates about WIPP, find out where candidates stood on the issue, or tell candidates they opposed the site. "In short," Colombo said, "the advertising campaign made WIPP an issue without solidifying and strengthening the opposition as effectively as it might have."[45]

Nevertheless, Southwest Research and CARD turned their focus to the New Mexico legislature, hoping to capitalize on the growing public sentiment. The previous year, a proposal to put WIPP to an immediate statewide vote had narrowly failed in the House, 36–34.[46] Organizers worked with groups around the state to gather 20,000 signatures on an anti-WIPP petition. But Southwest Research, short of cash, was able to play only a limited role in assisting in work for the 1979 legislative session. In addition, the group faced formidable opposition in the legislature. Carlsbad's Joe Gant chaired the Senate Conservation Committee, and eight of the panel's 14 members supported the project. Similarly, in the House, Carlsbad's Jack Skinner took over as chairman of the Energy and Natural Resources Committee, where eight of the 11 members backed WIPP. Lawmakers representing Los Alamos and Sandia laboratories were supportive, as were members from the uranium-mining country in northwestern New Mexico. They saw WIPP as the necessary closure of the "back end" of the nuclear fuel cycle that would sustain their industry.[47]

Of the half-dozen pieces of WIPP-related legislation that were introduced during the session, two bills drew the most attention. The first, introduced in the House, was called the "Conditions Bill" because it set forth a number of requirements that had to be met before construction on WIPP could begin. The requirements included a state decision on WIPP, licensing by the Nuclear Regulatory Commission (NRC), and payments to the state for all natural resource royalties that would be lost at the site if WIPP was built. The second was the "Referendum Bill," a Senate measure that provided for a statewide referendum on permitting nuclear waste disposal in New Mexico. In the end, after a great deal of negotiation and horse trading, the Senate bill was defeated while the House bill ended up being watered down far beyond what critics had wanted.[48]

Colombo cited several reasons why environmentalists failed to win passage of either bill in its original form. Perhaps most significantly, he noted, they were competing against entrenched energy and government

interests. Three legislators with nuclear backgrounds—John Rodgers and
Vernon Kerr from Los Alamos and Skinner, whose Carlsbad firm handled
uranium equipment development—played important roles, as colleagues
took their views seriously. Kerr, a chemist at Los Alamos, had already
questioned the accuracy of the information that Southwest Research had
disseminated on WIPP and faulted the group's scientific credentials. "They
do not do any bench-type research," he said dismissively at a July 1978
hearing. Kerr showed little interest in meeting environmentalists' demands.
He described the situation with WIPP as "Orwellian," with opponents of
nuclear energy "saying ignorance is knowledge; they are also saying misin-
formation is truth."[49]

Lawmakers like Kerr found allies who were willing to fight it out in the
court of public opinion. Among the most pugnacious was John Dendahl,
who was president of Eberline Instrument Corporation, a manufacturer
of radiation-monitoring equipment and a well-known figure in state
Republican circles. The outspoken Dendahl became chairman of a group
called New Mexicans for Jobs & Energy, a coalition of labor, business,
and professional interests that lobbied for energy-related development
in the state. "Some voices in New Mexico argue that our electricity is
not being generated by nuclear power plants, so we should not bear
whatever environmental strain may be caused by nuclear fuel cycle
activity," he told a congressional committee in 1979. "Missing from that line of
reasoning, however, is any mention of the myriad products we all enjoy,
the creation of which creates environmental strain elsewhere—chemical
products, products made from plastic, steel, paper, aluminum, rubber
and so on."[50] Dendahl began writing letters to the editors of newspapers
across the state when articles or editorials critical of WIPP appeared. As
he saw it, the Energy Department's unwillingness to make public information
a priority made his role necessary. "They didn't want to communicate
honestly with the public," he said in 1999. "That just fed the antis."[51]

Another problem Colombo discovered with the anti-WIPP forces'
strategy in the 1979 legislature was its inflexibility. After they realized early
in the legislative session that victory was unlikely, the critics decided to
emphasize private lobbying, something Colombo regarded as a tactical
mistake. It did not help that they lacked a well-informed and firmly

committed leader among lawmakers to champion their cause. The principal sponsors of their bills were too preoccupied with other legislation and more amenable to compromise in order to move the WIPP bills along. In the end, Colombo questioned whether the groups should have bothered with the state government at all. "Anti-WIPP groups were not mobilized on the scale necessary to win in the Legislature," he said in his doctoral thesis. "The pro-WIPP forces had the preponderance of political power to control these decisions."[52]

Three Mile Island Solidifies Opposition

The political push to continue WIPP left Montague disenchanted. He moved to New Jersey in 1979 and eventually relocated to Annapolis, Maryland, to run another environmental research group, although he remained on Southwest Research's board of directors. "In retrospect, I think, the day they began digging in the ground, the fight was over," he said in 1999. "Once they were allowed to begin to dig, it was silly to think they were ever going to back off that project. But I don't think we realized that at the time."[53]

Other critics of WIPP, however, found reason for optimism in the late 1970s. Nationally, the tides of public opinion were turning in their favor. In August 1978, the problems at Love Canal near Buffalo, New York, came to light, igniting public outrage over illegal waste dumping. President Carter declared a state of emergency as hundreds of residents fled their homes on learning that 20,000 tons of toxic chemicals were buried under them. Then in March 1979, an even more dramatic incident provoked a tectonic shift in thinking about nuclear energy: Three Mile Island. Before the mishap at the plant near Harrisburg, Pennsylvania—the worst disaster in U.S. nuclear history—a majority of the American public apparently believed that nuclear generation of electricity rested on a proven and fundamentally safe technology. But the effect on public opinion produced a shift from an almost 2-to-1 margin of support before the accident to a roughly even split between supporters and opponents of nuclear power immediately afterward.[54] A member of the NRC acknowledged that the incident shifted the burden of proof from the nuclear opposition to the industry and its regulators by demonstrating that the catastrophic

Figure 10: *Don Hancock of Albuquerque's Southwest Research and Information Center (Southwest Research and Information Center photo).*

potential of nuclear power was more than just the opposition of a handful of fanatics.[55]

Among those mindful of the growing loss of public confidence was Don Hancock, a 31-year-old staffer at Southwest Research at the time of the Three Mile Island accident. Hancock had joined the organization in 1975 after working with a Methodist peace group in Washington. A native of Indiana, Hancock had received his degree in political science from DePauw University five years earlier; he wanted to work for social change, so he traveled to Albuquerque and asked Montague if he needed a volunteer. "Three Mile Island was an example that proved what we were talking about: radiation was dangerous, accidents could happen," Hancock said in 1997. "It helped exacerbate the opposition. The people who were already concerned became even more concerned."[56]

Hancock embodied many of the traits of his generation of 1960s activists. He was passionate, smart, forceful, and idealistic. He also was innately suspicious about taking the government at its word. Consequently, he had little regard for diplomacy, and his tart remarks in the news media and at public forums won him widespread enmity among project officials. Many of them questioned his credentials, often pointing out to reporters that Hancock was not a scientist. Hancock fervently agreed with Anderson that WIPP was driven not by science but politics. Rather than give the department the benefit of the doubt to investigate whether the project was safe, he said, "I would argue that DOE is guilty until proven innocent."[57]

Environmentalists were impressed by the dedication with which Hancock approached his mission as a critic. He put in long hours seven days a week, earning a paltry salary. He read voraciously, accumulating a familiarity with the project's legal, technical, and political details that would prove unsurpassed among environmentalists and developed contacts among scientists, bureaucrats, and others. His sheer breadth of knowledge would, over time, effectively refute continued protests about his lack of credentials. "I think [opposing WIPP] is a religious thing with him," Montague said. "He's basically taken vows of poverty in his personal life and devoted his life to this."[58]

Hancock eventually emerged as the Ralph Nader of radioactive waste disposal. Just as Nader passionately attacked corporate America and built a loyal following, Hancock did more to contribute to the climate of scientific doubt around WIPP than any other figure. Many environmentalists, locally and nationally, took their cues from him. In terms of longevity on the project, his only rival was Weart, who continued to defend the science being done on the project while lamenting the public backlash. "After Three Mile Island, the environmentalists said, 'We got you where we want you. This is the end,'" he said. "And in a sense, they were right." Weart acknowledged that critics such as Hancock had a noticeable impact. "We didn't fully understand the degree of public opposition we would get," he recalled. "They've been fairly effective in holding our feet to the fire and having us cross every T, dot every I, provide absolute assurances in terms of technical information when the predominate weight of technical judgments would have said, 'Yeah, it's all right, go ahead.' We ended up having to produce high-quantity data justifying every one of our positions." Such an approach, he contended, was not being done in the name of safety "but to further another agenda."[59]

Hancock and others, however, often disagreed that the data being produced on WIPP was of high quality. As the 1980s began, they spent time poking holes in a revised environmental impact study that had been drafted for the site. The study called for drilling two shafts to a depth of 2,150 feet and building underground experimental rooms to test such issues as the corrosion of nuclear waste containers and the response of the salt beds to excessive heat. The department ruled out the possibility of leaving the waste in Idaho because "in the long term, some natural events and human intrusion that might produce large exposures are probable." At the same time, it maintained that salt was "the best understood of all candidate geologic media with respect to its possible use as a waste repository."[60]

The critics cited several problems with the study. They said the range of alternatives examined was too narrow, with not enough emphasis given to experiments without any waste. They also said it ignored the effects of future climatic changes on salt dissolution rates, one of Anderson's major concerns. And they said the report was sadly lacking in details on what would be done with the waste if it needed to be retrieved. "In general, there seem to be many important unsettled issues regarding WIPP," Montague said in presenting Southwest Research's objections.[61]

Most New Mexico citizens, for their part, appeared inclined to side with the environmentalists. Three days of public hearings in Carlsbad, Santa Fe, and Albuquerque drew more than 1,500 people. Many expressed frustration that the process was skewed toward continuing with WIPP in the absence of a consensus over its merits.[62] By September 1980, a new Zia Research/ KOAT-TV poll showed that statewide support among those polled had remained steady at 26 percent, while opposition had jumped to 63 percent from 54 percent two years earlier.[63] As environmentalists doggedly continued to raise questions, the Energy Department's plans for WIPP also were increasingly troubling to Congress and New Mexico's state government. Each would inject itself into the political process and register substantive changes in the direction of that process, setting it on a different course for the future—but without succeeding in shutting it down.

3

"An Equal Partner"

1978–81: CONGRESS, NEW MEXICO SEEK CONTROL

As the 1970s drew to a close, Melvin Price and Jeff Bingaman were two very different politicians at opposite points in their careers. Price, an elderly Democratic congressman from East St. Louis, Illinois, was finishing a tenure on Capitol Hill that began when Franklin D. Roosevelt was president. He was chairman of the House Armed Services Committee, a position that gave him a platform for his belief that the military needed more money than most members of his party thought necessary. A former sportswriter who never finished college, Price was a halting public speaker who was more comfortable talking about the St. Louis Cardinals than the defense budget. He was a product of the grimy industrial region across the Mississippi River from St. Louis and one of Congress's strongest advocates of nuclear power. He was a former chairman of the Joint Committee on Atomic Energy and in 1957 had co-sponsored the Price-Anderson Act, a landmark measure that enabled the nascent commercial nuclear industry to establish itself by limiting its financial liability in the event of accidents.[1]

Bingaman, meanwhile, was a young Democrat on the rise. A serious-minded pragmatist who was longer on intellect than charisma, his father was a college professor in Silver City, situated in southwestern New Mexico's mining country, and his uncle was a trusted confidant of former senator Clinton Anderson. While attending Stanford Law School in the mid-1960s, Bingaman worked on Robert F. Kennedy's 1968 presidential campaign. On returning to his home state, he joined a politically connected law firm in Santa Fe before successfully running for state attorney general in 1978 at age 35, his first bid for elective office.[2]

In separate but parallel ways, Price and Bingaman helped mold the future of WIPP. By challenging the Energy Department and taking the lead on efforts to carve out formal roles for Congress and New Mexico, their actions allowed the project to proceed at a critical time in its early history. The results filled a policy vacuum created by an executive branch that had failed to articulate a clear and convincing scope for WIPP and that had stumbled badly in its efforts to forge a consensus with Capitol Hill and New Mexico's government. In Price's case, the action took the form of legislation that both blocked the department from storing commercial high-level wastes and reversed President Jimmy Carter's subsequent proposal to cancel the plant. Later, Bingaman filed a lawsuit against the agency stemming from New Mexico officials' unhappiness over the perceived unresponsiveness to their concerns. The litigation led to a settlement addressing transportation, emergency preparedness, and other issues.

The two men's initiatives took place as an early political consensus emerged on the project, supplanting a public consensus that remained elusive in New Mexico. Much to the dismay of critics, many politicians began adopting the view that enough time and energy had been devoted to the New Mexico site for it to be used. As one powerful senator, Democrat Henry "Scoop" Jackson of Washington state, declared in 1980: "We've worked too long and too hard to resolve the differences over WIPP not to go ahead with it."[3]

Price's and Bingaman's efforts also came during a period in which the Energy Department no longer controlled the nuclear waste agenda. Spurred by the energy crisis, the Three Mile Island nuclear plant catastrophe, and other events, waste storage became a hotly debated national issue during Carter's presidency, with the White House and Congress active in setting policy. Like so many later developments involving the New Mexico project, external forces that created conflict influenced the process, thwarting efforts at consensus.

That pattern was established in 1976, when California enacted a statute that linked continued development of nuclear power within the state to the federal demonstration of a permanent waste disposal technology.[4] Other states also were vocal about expressing objections. In Illinois, the state's attorney general fought efforts to ship low-level wastes to a temporary burial site in Sheffield. In New York, state attorneys battled a private company over who would be responsible for decontaminating a spent fuel site near Buffalo. Michigan governor William Milliken told the government that no exploring for commercial waste sites would be done in his state; Louisiana legislators took a similar stand. Meanwhile commercial power plants were filling up "temporary" storage pools with high-level spent fuel at a rate of 150 tons a month. As one industry official acknowledged: "Just to have a [storage] facility in place would put everybody's mind a little more at ease."[5]

The situation led Energy Research and Development Administration (ERDA) officials to hastily begin looking at their options. Two months after California's action, ERDA announced it would search in 36 states for six sites that could hold high-level commercial spent fuel, with New Mexico excluded because of its status hosting WIPP. That idea failed to turn up any candidates as states quickly became aware of how little power they might wield. Faced with objections, the agency abandoned its search and began looking seriously at WIPP as a commercial-waste candidate. In a significant policy shift, project manager Delacroix Davis announced in January 1977 that the New Mexico site would be designed to be licensable by the Nuclear Regulatory Commission (NRC), the federal overseer of civilian nuclear wastes. Although the regulatory commission would have been required to license WIPP if high-level military wastes were ever brought, Davis made clear that if geological studies indicated it was possible, "consideration would obviously be given to making it a commercial [high-level] site."[6] Such optimistic statements showed questionable political judgment.

That same year, after the formation of the Energy Department, a task force led by Deputy Secretary John Deutch assessed the storage situation. When it completed its report in February 1978, the task force called for disposing of high-level wastes in a permanent repository. Its members also argued for further research and development in underground salt beds, as well as a technical demonstration of the feasibility of putting commercial spent fuel at a monitored underground site. Although WIPP met those requirements, the department's plans at the time limited the site to receiving only transuranic, or medium-level, wastes from defense bomb factories. Energy officials said the change resulted from Carter's decision to prohibit the reprocessing of spent fuel assemblies. They wanted to prevent the worldwide spread of nuclear-weapons-grade materials that could fall into the hands of terrorists or hostile foreign governments.[7]

Addressing New Mexico's Dissatisfaction

Until this time, New Mexico politicians followed Governor Bruce King's approach of tacitly accepting studies of WIPP. When the project was purely military in nature, the state sought no active role in decision making. Jerry Apodaca, the Democratic governor who served between King's two terms, declined a request in 1977 from the chairman of the state Environmental Improvement Board to exercise veto power over the project. "I have believed that the proper position for me to take is one of concerned neutrality, awaiting the evidence before taking a firm position," Apodaca said in a letter to the board.[8]

Eventually, though, New Mexico politicians began to join other states in becoming alarmed about what was under discussion. Suddenly, it seemed, a demonstration project for materials from the nation's bomb factories that could arguably be justified as helping the U.S. defense effort could, instead, become the national garbage dump for commercial by-products from Maine to Mississippi—but not their own state. That, they realized, was considerably more difficult to defend politically. It hardly helped the Energy Department's cause that a private company, Chem-Nuclear Systems, Inc., proposed to build a separate low-level waste disposal site near Cimarron, a struggling town in New Mexico's rural northeast. The company abandoned the idea in 1978 after encountering substantial environmental opposition.[9]

Concerned about WIPP, members of New Mexico's congressional delegation sent Energy Secretary James Schlesinger a letter in November 1977 warning they would not readily accept any major restructuring of the plant's functions. The letter expressed disappointment about the potential inclusion of high-level waste—not because it might be more hazardous than the military waste but because the state had not been consulted. "It seems likely that DOE may soon give undue credence to the notion that the WIPP site is the most suitable location in this country to receive both

Figure 11: *New Mexico senator Pete Domenici* (Congressional Quarterly *photo*).

domestic and foreign radioactive waste materials from commercial reactors. . . . We want to register our conviction that this position cannot reasonably be taken by the federal government without the informed concurrence of the people of New Mexico," the lawmakers wrote.[10]

Among the troubled politicians was the state's senior senator, Pete Domenici of Albuquerque. To Domenici, who was running for reelection in November 1978, talk of emphasizing commercial wastes over military materials at WIPP was "inappropriate and premature."[11] Domenici did not take kindly to being challenged. The son of an Italian immigrant, he was a former minor league baseball pitcher—known for his wicked fastball—and an intensely driven man. Fiscally conservative yet moderate on many social issues, he fit the mold of western Republicans who believed in balancing environmental protection with economic development. Although usually mild mannered, he could bristle at anyone who questioned his actions. He had been elected five years earlier, at age 40, after having served on the Albuquerque City Commission and making an unsuccessful run at King in the governor's race. Domenici was several years from being thrust into the national spotlight as the deficit-conscious chairman of the Senate Budget Committee. In his initial years on Capitol Hill, his reputation was that of a hardworking legislative dabbler, with an eye toward protecting the federal interests of his state.

A year earlier, Domenici and fellow New Mexico Republican Senator Harrison Schmitt, a former Apollo astronaut, joined in voting against a failed legislative attempt to give states the authority to reject proposals by the federal government to locate nuclear waste sites within their jurisdictions. Both men contended that the measure, sponsored by Democratic senators Floyd Haskell of Colorado and George McGovern of South Dakota, was too broadly drawn and would have had the effect of killing off the commercial nuclear power industry. Domenici, however, said he remained open to some limited form of veto power.[12]

Aware of the need to pacify Domenici, Schlesinger met with the state's congressional delegation in February 1978. The secretary tendered what the lawmakers regarded as a startling offer: New Mexico would have a chance to veto WIPP. In the New Mexico legislature, House Majority Leader David Salman had introduced a bill to that effect that would later fail by just three votes. Schlesinger's deputy, John O'Leary, later explained that making such a concession to New Mexico amounted to the federal government's recognition that a repository could not be built if the host state was intent on fighting it. O'Leary had particular experience with the officials involved; he had served as New Mexico's natural resources secretary under Apodaca. "When you think of all the things a determined state can do, it's no contest," he said, citing the authority of states over lands, highways, and other issues. Although he acknowledged that the federal courts could overrule the state's authority, O'Leary argued the practical effect of such opposition was that it would prevent a site from ever being built.[13] Such thinking may have seemed logical but later would not make much political sense.

New Mexico officials reacted cautiously to Schlesinger's offer. Domenici, realizing the promise was at the secretary's discretion and that his successor would not have to observe it, vowed to have Congress pass new state veto legislation. He introduced his own bill designed to provide states with full consulting authority, the power to veto a disposal facility, and the right of the state legislature to determine for themselves the process to ensure full public participation in the siting process. But the first-term senator had no success. The Carter administration saw it as an intrusion on its prerogatives, citing constitutional grounds.[14] Others in Congress also were skeptical. Washington's Jackson, for one, believed that WIPP as an unlicensed facility for military waste could end up indirectly helping the commercial nuclear power program, providing some early data on the problems of geologic disposal.[15]

In any event, Schlesinger's veto offer did little to clarify the circumstances surrounding New Mexico's role in nuclear waste disposal. Before long, the department appeared to retreat from its initial promise, landing it in more trouble. O'Leary began discussing the state's "right to concurrence"— something more equivocal than veto power, since it did not address how to resolve a situation in which concurrence was unachievable. In July, O'Leary renewed the promise of a veto—calling it "from the heart"— but at a briefing for state officials in October, he again characterized the state's role as one of concurrence.[16] The appearance of a retraction did considerable damage to the department's credibility. "The final decision rests with the state," Schmitt said. "I've detected Schlesinger and O'Leary backing off from the original commitment that Schlesinger made."[17] State Energy and Minerals secretary Nick Franklin agreed, declaring that "the federal government must make New Mexico an equal partner in reviewing the proposed WIPP project, or face the possibility that the plant will never be built."[18] Such rhetoric was a far cry from the open-minded, wait-and-see approach of a few years earlier.

Because of its failure to articulate a consistent vision and state oversight role for WIPP, the department went from dealing with a once cooperative state government to a hostile one. Another opportunity for achieving consensus was lost. The department, given its history of doing as it pleased when it came to nuclear weapons, was unprepared in the fine art of accommodating local concerns. New Mexico's combative stand troubled O'Leary, a believer in the idea that the federal government had an obligation to help nuclear power establish itself as a viable energy source. He had hoped, perhaps naively, that states would put the national interest above their own. At a public hearing in April 1978, he lamented that only one

new order for a new U.S. nuclear plant had been placed in the previous
two years, with 14 cancellations of existing orders.[19] "He had a strong
belief that the United States had to prove its mettle to meet the energy
needs of the nation by ensuring we would do something with the waste,"
said his former wife, Hazel O'Leary, who would later become secretary
of Energy. "In the years I knew him, he always bemoaned the lack of
public acceptance for nuclear power."[20]

New Mexico's WIPP Watchdog Forms

As the potential magnitude of WIPP became apparent to New Mexico
officials, they began developing responses that conveyed the seriousness
with which they now took waste storage. In the 1979 state legislature, a law
was enacted to establish the interim Radioactive and Hazardous Materials
Committee of lawmakers to perform an oversight role over WIPP and related
waste issues. With Carlsbad legislators such as Joe Gant already holding
powerful positions on other committees, the plant's boosters had little
trouble ensuring the committee maintained a pro-WIPP tilt. Lawmakers also set up the state Radioactive Waste Consultation Task Force, an
executive branch group of three cabinet secretaries, to serve as a political
and informational liaison between the state and Energy Department.[21]

Far more important, though, was the decision in 1978 to form an expert
scientific group to review technical work on WIPP. With encouragement
from O'Leary, the Environmental Evaluation Group (EEG) was established
to assist the state Environmental Improvement Division in looking at
the environmental, health, and safety research on the plant. The motives
of O'Leary and others at Energy were not entirely altruistic—they realized
they risked losing the state support they had taken for granted if they did
not allow for external scrutiny and potential validation of their efforts.
The governor's technical excellence subcommittee that had been formed
under King was a group of part-time volunteers unable to devote much
time and attention to the project.

The formation of a full-time watchdog group was a significant step.
As author Luther Carter concluded, "It contributed to defining WIPP as
an issue that could be dealt with on its technical as well as its political
merits."[22] From the outset, it was understood that the evaluation group's

Figure 12: *Environmental Evaluation Group director Robert Neill (Environmental Evaluation Group photo).*

ability to function hinged on its independence, with neither the Energy Department nor the state interfering with its judgments. The money that the department agreed to provide was used to hire full-time professional staff and outside consultants. One of the first consultants brought aboard was the University of New Mexico's Roger Anderson, whose views on salt dissolution had caused such a stir.

The man chosen to guide the monitoring process was Robert Neill, a loquacious New Jersey native who had been an associate director of the U.S. Bureau of Radiological Health. His selection as the group's director triggered suspicion from those on both sides of the debate— environmentalists accused him of being a pawn of the nuclear industry, while supporters feared he was bent on killing the project. The polarized situation led Neill to steer a careful middle path. From the outset, he kept his group focused on scientific issues rather than the murkier legal or socioeconomic concerns. "WIPP should be given a fair hearing," he told a state legislative committee in January 1979. "It shouldn't be approved before all the facts are available, nor should it be disapproved before there is a rational basis to judge."[23]

With politicians such as Domenici in its corner, the evaluation group's political viability remained solid. But over time, Neill and his organization would be a source of anger and disappointment to those on both sides of the WIPP debate. Environmental groups praised its willingness to join them in challenging the department but bemoaned its lack of power to enforce its recommendations. In 1984, a postgraduate student at the University of Oregon wrote a research paper that implicated the group in a "cover-up" with the Energy Department for failing to state its concerns because of fears it might be put out of existence—a charge Neill dismissed.[24] Meanwhile, WIPP proponents in Carlsbad and elsewhere regularly accused the group of devising or exaggerating problems solely to perpetuate its function. Neill usually shouldered such concerns in stride, using an irreverent sense of humor to disarm detractors. In the years to follow, he often wisecracked about the department's continued inability to meet its scheduled opening dates for bringing waste. "If WIPP was a musical," he liked to say, "it would be called 'Promises, Promises.'"[25]

In contrast, one area that Neill took quite seriously was fighting the public perception that the transuranic wastes from weapons plants headed to the site were relatively harmless "low-level" materials. Much to his annoyance, media accounts often glossed over the definition of "transuranic" and used the low-level description. Neill reminded reporters and remarked in speeches that the effects of plutonium in transuranic waste were remarkably long-lived. "These materials are not merely rags and paper and plastics and metals," he said. "That's deliberately misleading."[26]

Neill's group did not always have an easy time making its case. A succession of Energy officials denied the group the access it sought to technical data and sometimes ignored its recommendations. Nevertheless, the evaluation group became a key external check on the project. In general, the group supported the purpose of WIPP but consistently took a more cautious approach than the department in urging further study of potential problems. Despite its inability to enforce its recommendations, it did prompt the department to alter its plans. After the "Lake WIPP" underground brine encounter in 1981, the EEG recommended building the repository to the south of the central shaft rather than the north, and the change was made. Investigating Anderson's claims about salt dissolution, the group recommended that the department conduct deep-drilling tests of a possible source of dissolution two miles north of the site. It eventually decided that no dissolution occurred there. By 1983, the group was sufficiently satisfied with the geologic and hydrologic issues to conclude the project "has been characterized in sufficient detail to warrant confidence in the safety of the site."[27]

Price Takes on the Energy Department

While President Carter's administration helped establish an external review process on nuclear waste disposal, it also sought to achieve an internal consensus. The administration had formed an interagency group that expanded on the efforts of the initial Energy Department task force by including representatives from 14 federal agencies. In a draft report in October 1978, the task force recommended the construction of a licensed repository that would demonstrate the safe disposal of 1,000 high-level commercial spent fuel assemblies. To O'Leary, the disposal demonstration was important because it would show lawmakers and the public that the department could satisfy the California-imposed moratorium. But even though the review group's report clearly seemed to point to WIPP, it never actually mentioned the project.[28]

The lack of any mention was the executive branch's recognition that any final decision on WIPP's fate would need to be made on Capitol Hill. Given Price's passionate support of nuclear issues, it was natural for Energy officials to assume they would receive friendly treatment in his committee. But Price became a formidable opponent of the agency, for reasons quite separate from New Mexico's. His concerns stemmed from the Nuclear Regulatory Commission's (NRC's) potential involvement in licensing the site as a high-level facility. Involving the NRC, he feared, could only lead to further delays and jeopardize the project—and in so doing, call into question the viability of nuclear power. He had political turf to protect. A repository designated solely for military waste would remain under the control of the Armed Services committees, whereas one that held commercial waste would fall under the jurisdiction of other committees, including House Interior and Senate Energy.

Thus emboldened, Price's Armed Services Committee refused in May 1978 to authorize money for the department to seek an NRC license, then prevented the environmentally conscious chairman of the House Interior Committee, Arizona's Morris Udall, from restoring the money. Domenici and other sympathetic senators eventually reinstated language to ensure the NRC's involvement, but the compromise conference agreement between the House and Senate called for merely studying the impact of licensing while eliminating three-quarters of the project's funding.[29]

The growing uncertainty over whether WIPP should be used for military or commercial wastes came to a head in December. Schlesinger, in an effort to silence objections over the licensing of any military site, proposed that the site be purely commercial. But Price, who was interested in maintaining jurisdiction over the project as a defense-related matter, tartly responded that if Schlesinger had that approach in mind, he should find another committee to get a new authorization for WIPP.[30] A month later, Price put the plant's status in limbo by omitting from the fiscal 1980 authorization bill all funding for the project—demanding, in essence, that the plant be exclusively military or it would not exist at all. Energy officials said the impasse meant no further action could be taken until Carter reviewed a series of waste management options.[31]

The Senate Armed Services Committee, which had taken a softer position on the Energy Department's plans, did include money for WIPP in its authorization bill. But the Senate also contained no provision for commercial waste storage—a move that effectively shut the door on putting materials for nuclear power plants at the repository. Faced with such a divisive situation, the department announced in May it would abandon its earlier proposal to store 1,000 spent fuel assemblies at WIPP and decided to return the plant to its original scope as a storage site for defense-generated wastes.[32] Such a concession normally might have paved the way for a consensus among the two branches of the federal government. But New Mexico officials had become distrustful enough of the Energy Department to take a hard-line negotiating approach. The state's attorney general, former King aide Toney Anaya, called for an immediate halt to work on WIPP and charged the department with "purposely misleading the people of New Mexico" in its determination to establish the site.[33] The result left ·

Figure 13: *House Armed Services Committee Chairman Melvin Price of Illinois* (Congressional Quarterly *photo*).

Congress in the unique position that the department had earlier faced— how to incorporate a state entity into what had been an exclusively federal decision-making process. The House bill gave New Mexico the authority only to review and comment, not to veto, forcing the state into essentially a bystander's role.

Making matters worse for state officials was O'Leary's resignation from the department in July 1979 following Carter's decision to fire Schlesinger. The sudden departure of the department's second in command "has apparently left Gov. Bruce King dangling without a contact in Washington to help represent the state's position on WIPP," the *Albuquerque Journal* lamented in an editorial.[34] Schlesinger's successor, Charles Duncan, reaffirmed the department's earlier promise of "consultation and concurrence" but tried to persuade King to instead accept a role that was more in line with consultation and review—a reduced role that was acceptable to Price. The Armed Services chairman disdained state concurrence, worrying that if New Mexico flexed its muscle and bowed to environmentalists' pressures, the entire project could go down the drain. New Mexico lawmakers, however, found such an offer an affront to state prerogatives.[35]

The impasse triggered an exchange of letters between King and Price. The governor tried to argue that New Mexico's historic contributions to nuclear weapons development and research meant the federal government owed it the final say over WIPP's outcome. Price's legal counsel on Armed Services, Adam Klein, accused King of failing to understand the federal system. The state's position baffled Klein; why, he wondered, would it carp so much about the project while seeming to tacitly accept it? In October, with both sides refusing to budge, the governor and the congressman met face-to-face. Neither man backed down, although Price clearly had the upper hand. King was in a delicate spot: he could not take too tough a bargaining position when the Armed Services chairman was in a position to retaliate against the economically vital Energy Department laboratories in the state, Sandia and Los Alamos. For the same reason, Harold Runnels, a member of Armed Services whose district included the project, could not risk alienating his committee's chairman.[36]

But when the authorization bill came to the House floor for a vote in November, Price made a surprising turnaround that became one of the defining moments in WIPP's history. He agreed to reverse his committee's recommendation and include funding for WIPP but with no state participation in decision making. Without consulting New Mexico officials, he offered an amendment stipulating that the plant be constructed solely as a defense facility and that any state veto be prohibited. "Unfortunately, the WIPP project has become embroiled in bureaucratic politics within the current administration and in the politics of the state of New Mexico," Price said in a speech. "I think that even those in the highest level of management in the Department of Energy will admit that the project has been mishandled by the department." His amendment, he said, "will simply return the project to the same status that it was when it was first presented to our committee." The inclusion of the prohibition of state veto language, he added, reflected the fact that a state government could not thwart the federal government's will. "I do not believe that any member of this body would agree to the expenditure of federal funds for the purpose of constructing any kind of a federal project which, after its completion, could not be used as a result of political action within a state," he said.[37]

In a show of the deference that members of Congress accord a powerful committee chairman, the amendment passed without a roll-call vote, a procedure usually reserved for noncontroversial actions and one not usually employed on issues as divisive as nuclear waste storage. New Mexico's two House members, Runnels and Republican Manuel Lujan, were in no position to object—both were attending a dinner in Albuquerque and missed the vote. Even so, Price's influence over other lawmakers was such that Klein said later that the presence of the two New Mexicans would not have made any difference.[38]

Back in Carlsbad, supporters rejoiced over the House's action. A letter to the editor of the *Current-Argus* had suggested, only somewhat tongue in cheek, that a WIPP "weather vane" be placed atop the newspaper building to indicate on a daily basis whether the project was alive or dead.[39] "Without Mel Price, this would have never moved," Louis Whitlock said.[40] But Southwest Research's Don Hancock lamented the degree to which the political process allowed Price to single-handedly prevent the NRC from overseeing any aspect of WIPP, continuing a pattern of removing the public from the decision-making equation. "The takeaway of NRC licensing was done by one man in two or three minutes," he said. "Nobody had a chance to weigh in on that kind of legislation." Price's action, at least, blocked any dangerously radioactive spent fuel from coming to New Mexico. "Mel Price gave us WIPP," Hancock said, "but he gave us WIPP without high-level waste."[41]

Price's stealthy action set the stage for a turbulent conference with the Senate. Thanks in part to Domenici, the Senate authorization bill continued to allow state concurrence authority. But Domenici was not among the senators chosen to negotiate the bill with the House. Instead the Senate conferees included Washington's Jackson, Colorado's Gary Hart, and South Carolina's Strom Thurmond—all representatives of states with defense facilities ready to send waste to New Mexico. After a series of difficult negotiations in which Price later said the idea of discontinuing WIPP altogether was seriously discussed, the final conference report adopted most of the language of the House amendment. Price remained in favor of moving ahead with WIPP as well as vehemently opposing veto power for the state. But as a concession, he agreed to language setting forth a framework for a new concept, "consultation and cooperation." The conference report gave the Energy Department and the state until the following September to work out the details.[42]

In their report, the conferees recognized the possibility that the secretary of Energy and the state might not be able to negotiate a mutually satisfactory agreement. "However," the report said, "in view of the long history of cooperation by the state with the federal government in atomic energy matters, and in view of the national significance of the WIPP project, the conferees fully expect such an agreement can be successfully negotiated. In this spirit of cooperation, no specific legislative remedy has been included to resolve a situation where there is no agreement."[43] Such abiding faith in federal-state relations would later prove overly optimistic.

Carter Cancels, Congress Revives

The congressional action set the stage for the largest conflict within the Carter administration over WIPP. As the 1980s began, the president made clear his distaste for Capitol Hill's decision, issuing a statement declaring that the future of waste disposal "ought to be resolved only in the context of an overall waste management policy." The president promised "a comprehensive statement on the management of nuclear wastes in the near future."[44] Rumors began circulating in Washington that Carter would scrap the project entirely in favor of a recommendation from the interagency task force that several potential sites be evaluated for commercial and defense wastes. Four weeks later, Carter proved the rumors true. "The administration has decided that the Waste Isolation Pilot Plant (WIPP) should be cancelled, and that defense waste previously intended for disposal in the WIPP facility should be placed instead in the first commercial waste disposal facility," an Energy budget report said.[45] In a subsequent message to Congress, Carter explained that Price's action to eliminate the NRC licensing requirement violated his recently completed waste policy, which included NRC licensing as a prerequisite for all permanent geologic repositories. The president asked Congress to rescind all funds for WIPP, although he said the site would continue to be evaluated along with four or five others.[46]

Administration officials said one reason for the decision was the poor coordination between federal and state agencies. They noted that WIPP was begun without consulting either the regulatory commission or the Interior Department, and little regard was given to New Mexico's ensuing objections. "We want to end that old way of doing things," one official told *The Washington Post*. "It's time to start anew."[47] Such a statement reflected the realization within the executive branch that Congress had seized control over setting waste policy. Environmentalists, however, scoffed that the Energy Department was incapable of such a move. "This is just a smoke screen to get all the people in New Mexico who have fought this thing to lay back and forget about it," said Ken Clark of Albuquerque's Citizens for Alternatives to Radioactive Dumping.[48]

Sure enough, Price again demonstrated Congress's ability to dictate the debate over WIPP by ignoring Carter's request. The Armed Services chairman said he was "dismayed" to learn the administration planned to abandon WIPP, especially since it had support from Congress, the Energy Department, and New Mexico.[49] Three months later, Price's committee breathed new life into the project by authorizing $5 million to be spent on it for the 1981 fiscal year.[50] The Senate Armed Services Committee followed suit, arguing that the project's cancellation would "significantly hamper defense weapons activities." It, too, reiterated that WIPP should be built solely for military wastes.[51] In the end, Carter ended up signing an appropriations bill providing $15 million for preliminary construction and site preparation work and $5 million for operations.[52]

The successful efforts of Price and other lawmakers to keep WIPP alive despite the administration's attempts to kill it illustrated the degree to which congressional committees jealously guard their domain—and how turf battles and other political considerations can supersede science and other concerns. In particular, the political jockeying showed how siting decisions for defense-related wastes and commercial wastes do not rely on the same political grounds for their legitimacy. As one study of the Congress-New Mexico WIPP dispute concluded: "In making a military siting decision, [the Energy Department] legitimately represented a national interest, but in a commercial siting decision it merely represented one side of a set of competing regional interests."[53]

In an unintended way, the decision to keep WIPP as a military-only facility by Congress helped save it. As Luther Carter has noted, WIPP almost certainly could not have been licensed for heat-generating high-level waste or spent fuel in light of the subsequent waste-retrieval requirements of both the NRC and the Nuclear Waste Policy Act of 1982. Under NRC regulations, such wastes must be available for retrieval at any time during the first 50 years of a repository's life span. Given the relatively rapid rate at which the walls close in at WIPP over time, scientists have concluded it is unrealistic to expect that waste would be retrievable for a full half century.[54]

Consulting and Cooperating with New Mexico

With the election of President Ronald Reagan in November 1980, relations between the Energy Department and New Mexico took another turn for the worse. Department officials, emboldened by the congressional mandate to continue WIPP for burying defense-generated waste, turned their backs on state involvement. They announced plans to forge ahead with drilling work later that year, with construction to be completed by 1983. The action was consistent with the department's emphasis on schedule setting. State officials formally asked in December 1980 for an extra 45 days to challenge the environmental impact statement for the plant, which had been issued in October; the department refused.[55] At a news conference the next month, WIPP project manager Joseph McGough declared, "We don't need anything else from the state, legally or officially." All that was required, he said, was permission from the Bureau of Land Management to use the land and a "continued flow of money from Congress." The news surprised and dismayed King. "There isn't much the state can do about WIPP now," the governor said.[56]

But state attorney general Bingaman disagreed. He believed the department had not yet addressed the state's concerns about WIPP's technical merits and had not followed Congress's instructions to open a dialogue. He sent new Energy secretary James Edwards a sharply worded letter asking for "immediate action" to stop the implementation of its plans and warned a court challenge was possible. Bingaman said he doubted the department's "piecemeal approach" to WIPP "will lead to a rational, intelligent or safe solution to the radioactive waste disposal problem, either at the state or national level." At the heart of his objections was whether a decision had

Figure 14: New Mexico
attorney general Jeff Bingaman,
who was later elected senator
(Congressional Quarterly *photo*).

been made to proceed before all necessary tests were completed. He also was concerned about liability in the event of an accident involving radioactive materials and what would be done to address transportation on state roads. Environmentalists fell in behind Bingaman, pointing to language in the National Environmental Policy Act (NEPA) of 1969 requiring the environmental impact statement to discuss a defined purpose and proposed action and to assess all reasonable alternatives of the proposed action. "There's no consistency whatsoever, which makes it very difficult for the state, or the citizens, to understand what they plan to do," Hancock said.[57]

King deferred to Bingaman's legal reasoning and started publicly siding with the attorney general. In April, the New Mexico offices of the Energy Department and Bureau of Land Management signed a cooperative agreement that they said eliminated the last obstacle to allowing underground shafts to be sunk for in-depth exploratory work. Both the governor and the attorney general noted that they did not receive a copy of the agreement until the day after it had been signed. The move substantially strained what little credibility the Energy Department had left. "This just the most recent example of them plowing ahead with this project without any serious attention to the concerns of the state," Bingaman said. King was angry enough to say he "just as soon" preferred that WIPP not be located in New Mexico, and he asked Bingaman to look at legal action. Two weeks after their comments appeared in newspapers, heavy construction equipment was sent to WIPP to begin site preparations.[58]

Bingaman, by that time, had announced he would run for the Senate in an attempt to unseat Schmitt the following year. But the attorney general strongly denied that political concerns or an antinuclear agenda motivated him. Indeed, he said he had a "gut feeling" that the design of the plant could prove to be sound. Thus his state's-rights argument enabled him to draw support from those who were lukewarm toward WIPP, such as King, and those actively hostile to it, such as Hancock. Predictably, though, Bingaman had a more difficult time with the Carlsbad faction. Before filing the lawsuit, he flew to the city to meet with proponents. "They told me that if I went ahead, that would be the end of my political career," he recalled. "I tried to explain I wasn't trying to stop the project . . . and I told them I was going ahead."[59]

The suit that Bingaman filed in U.S. District Court in May 1981 noted that the state had "no enforceable legal mechanism" for asserting its rights. "The initial phase of construction is termed as 'experimental,'" he said in a statement, "yet there is no assurance that New Mexico will be privy to the results of those experiments, or that our state will play a meaningful role in determining what activities should properly follow the experimentation." Among the laws violated, the suit said, were NEPA, the Federal Land Policy and Management Act of 1976 (FLPMA), and the WIPP authorization act approved by Congress in 1979. Albuquerque's Citizens for Alternatives to Radioactive Dumping already had cited the FLPMA violation in a separate lawsuit, contending that the law allowed public lands to be segregated from public use for only a two-year period. After that, the group said, a permanent withdrawal of the lands must be sought. The Interior Department had justified segregating the WIPP lands since 1976 by arguing that the Energy Department had submitted a new and significantly different application for each two-year period.

Also cited in the attorney general's lawsuit was the state "concurrence law" enacted by the 1979 legislature preventing the storage of nuclear waste until the state had signed off. Although Congress and the Energy Department's general counsel had determined New Mexico's veto power to be unconstitutional, Bingaman and other state officials continued to insist that they had been promised such authority. The suit said the Energy Department's actions unconstitutionally interfered with the state's sovereignty "in important and traditional areas of state governmental functions and responsibilities."[60]

Having enlisted a powerful ally in New Mexico's government, environmentalists were spoiling for a court fight. Key officials in the Energy Department's Albuquerque Operations Office also were prepared to slug it out, noting it would cost them $30,000 a day if work stopped.[61] But neither side would get to use the legal system. Energy secretary Edwards was a former South Carolina governor who knew how to placate a pragmatic deal maker such as King. The two men met in June to discuss New Mexico's concerns, and in a short time, a "consultation and cooperation" agreement was signed a week before U.S. District judge Juan Burciaga was to hear the state's case.

The department made two key concessions to New Mexico in exchange for having the lawsuit dropped. It agreed to consider and address the state's concerns before deciding to proceed with construction or bringing waste. It also formally acknowledged the state's right to seek judicial review of departmental actions regarding the project. The agreement required the department and Sandia National Laboratories to prepare 16 reports summarizing the results of all studies for experiments to ensure confidence in the stability of WIPP's underground storage rooms and passages. It set forth a framework with which the state could review the department's work at three separate periods, with the state allowed to return to court to stop the project if it remained dissatisfied at any of those junctures. Finally it obligated the department to make "a good-faith effort" to set up a state-federal task force that could address state concerns about road upgrades, accident liability, emergency preparedness, monitoring waste shipments, health studies of people living in the plant's vicinity, and the postoperation monitoring of the site.[62]

Bingaman was pleased with the arrangement. He noted that the state, at

long last, had "a binding and enforceable legal mechanism" for asserting
its rights. Filing the lawsuit, he said, was the only way to get the attention
of a department that had grown accustomed to doing things its own way
and that had come to realize, like Congress a year earlier, that states could
make trouble if not given input. "The result was a workable arrangement
between the federal government and the state that allowed the state to be
confident its concerns were being addressed at each step," he said in 1999.
"Obviously, it would have been better if we had been able to do it without
going to court, but that wasn't possible."[63]

With the state's concerns apparently addressed, an anxious Energy
Department wasted no time in moving on WIPP. Three days after Judge
Burciaga signed off on the agreement, it began drilling a 12-foot-wide shaft
to allow it to mine rooms 2,150 feet underground. A second six-foot-wide
shaft was drilled to help air circulate inside the new repository.[64]
Meanwhile, skepticism continued in New Mexico about the department's
ability to live up to its commitment. *Albuquerque Journal* cartoonist John
Trever depicted two Energy officials sitting in an office, reading
Bingaman's lawsuit. "Looks like we can still go ahead and do as we damn
well please," one official remarked. "They just want to be reminded of it
from time to time."[65]

Despite the dismay among environmentalists and others that a tougher
court-ordered action was not taken, the attorney general emerged with his
political fortunes unscathed. The issue became only a minor one in his
Senate campaign—he beat Jerry Apodaca in the Democratic primary,
then edged out Schmitt. Among the counties Bingaman carried was Eddy,
home of Carlsbad and WIPP.[66] Bingaman's experience as both an attorney
general and a senator left him with an understanding of the uniqueness
of interlocking federal-state-local government arrangements on issues as
delicate as nuclear waste storage. "It's a multifaceted process, a very drawn-out
process, and a very frustrating process," he said. "Having the executive
branch with its authority, the legislative branch with its authority, and
the courts involved with all of their authority, and all three interacting
and all three able to impact the course of events at many different stages,
it shows there's truth to the old adage that the wheels of government
grind exceedingly slow but fine."[67]

As the 1980s continued, politicians, environmentalists, and others would perfect the ability to take advantage of the complexities of such a political arrangement. In seeking to use WIPP to further their own agendas, they would end up preventing the Energy Department from continuing to follow its own agenda for WIPP. Their actions would create an environment in which outside entities sought openings so that they could influence the lumbering process of preparing the site as a permanent disposal facility. Whether they succeeded in their attempts all of the time remains debatable, but one thing would become clear: WIPP would not meet the department's goal of opening by the end of the decade.

4

"I'm Not in the Garbage Business Anymore"

1981–88: IDAHO, PLUTONIUM POKER, AND OPENING WIPP

It was October 1988, and Cecil Andrus was out of patience. Nearly two decades earlier, the Democratic governor of Idaho had helped extract a promise from the Atomic Energy Commission (AEC) that the plutonium-laced transuranic wastes on a temporary storage pad at Idaho National Engineering Laboratory would leave his state by 1980. Since that time, all that had happened was the arrival of more waste—an average of 55 shipments a year—and, he lamented, "there was no indication when it would end."[1] The commission's successor, the Energy Department, had foundered in its attempt to open the Waste Isolation Pilot Plant (WIPP), the expected burial site for the radioactive rubbish from Colorado's Rocky Flats plant that was being sent to Andrus's state. Key members of Congress had abandoned hope of passing legislation anytime soon to clear the way for shipments from Idaho to New Mexico. To Andrus, it was obvious something needed to be done.

The governor and his press secretary, Marc Johnson, traveled to Carlsbad, donned miner's helmets, and toured WIPP. By now, the plant was an elaborately engineered array of rooms and tunnels on which more than $700 million had been spent. A few months earlier, workers had unfurled a large banner across the complex proclaiming "Ready for Waste October '88." But the plant was not expected to open until the following year at the earliest. During the trip, Andrus chatted with his friend Louis Whitlock, now a New Mexico state senator. What would it take, the two men wondered, to get WIPP open? Andrus was pessimistic, believing no

Figure 15: *Waste drums in storage at Idaho National Engineering Laboratory on a temporary pad awaiting shipment to WIPP (Idaho National Engineering Laboratory photo).*

one in the federal government was willing to demonstrate any leadership. Flying back to Boise, he indicated what was on his mind. "On the airplane he looked over at me and said, 'I think I'm going to tell the federal government they can't bring any more waste into Idaho,'" Johnson recalled. The startled press secretary objected, but Andrus would not budge. "He knew all the arguments against it," Johnson said. "And he also could see five or six steps down the road."[2]

Down the road was a profound illustration of how, in the absence of a clearly defined agenda from the federal government, state governments could use the nuclear waste issue to advance their own parochial concerns. By trumpeting his refusal to accept any more waste, Andrus set in motion a sequence of events—the media labeled it a "crisis"—that would draw national attention and galvanize public sentiment regionally. It would place the Energy Department on the defensive, worried that other states might follow suit. It eventually would force the agency into a schedule-setting situation that provoked a difficult cycle of confrontation and crisis. Eventually it would win Andrus more money for environmental cleanup work at the Idaho laboratory. It was a classic case of politics colliding with science.

By this time, Andrus was skilled in the art of western environmental lawmaking. Charismatic and "admittedly bullheaded,"[3] he had a unique insight into understanding the federal government's limitations. After serving as governor from 1971 to 1977, he became President Carter's secretary of Interior.[4] When Carter left the White House, Andrus started a lucrative consulting business but discovered he missed the "fun" of politics. So he ran for reelection and moved back into the governor's office in 1987. "I had learned one basic lesson as a Cabinet secretary," he wrote in his 1998 autobiography. "The government in our nation's capital reacts only to crises. If we wanted action [on nuclear waste], we would have to create a crisis and force the Department of Energy to give us its attention."[5] It would be an important political lesson for the agency.

Nuclear Waste Policy Act and Its Aftermath

The seven-year series of events that led up to Andrus's display of bravado in the name of states' rights showed some progress on nuclear waste policy but also a continuing lack of political and public consensus. During the early 1980s, changing priorities within the executive branch still buffeted WIPP. The agency sought $80 million for the 1982 fiscal year to move forward with the project, but the budget cutters of new president Ronald Reagan's administration slashed that request almost in half. Later, after the department sought an increase to $125 million for the following fiscal year, the White House cut $98 million from that amount. Unlike the situation under Jimmy Carter, when Congress was forced to intervene to keep WIPP going, Energy secretary James B. Edwards successfully appealed the decision, and the full amount was restored.[6]

But Congress, and not the White House, continued to drive the debate. The new Energy Department became a ripe political target, with some Republicans calling for its abolition. The agency became beholden to a handful of lawmakers for its survival. Because the commercial high-level-waste issue had proved so controversial, both the House and the Senate Armed Services Committee were eager to keep defense and civilian waste management from becoming intertwined, citing concerns about the Nuclear Regulatory Commission (NRC). They also wished to avoid a repeat of New Mexico's defiant stand against accepting high-level waste. As a result, in the defense authorization bill for the 1981 fiscal year, lawmakers separated the two programs. The department opposed the move, contending it would result in a duplication of effort. But Congress remained adamant, and so in September 1981, the Energy Department transferred its defense waste programs from the Office of Nuclear Energy to the Office of Defense Programs. Commercially generated waste remained under the jurisdiction of the nuclear energy office.[7]

Reagan came out the following month with his waste management plan. The president instructed the secretary of Energy to work closely with industry and the states and to "proceed swiftly" toward finding a way to store and permanently dispose of high-level wastes, in essence continuing Carter's goals. Congress, having failed in 1980 to enact a comprehensive nuclear waste bill, renewed its determination to put its legislative stamp on the matter. But the issue of commercial versus military nuclear waste remained a sticking point, as did the "not in my backyard" opposition of many lawmakers.

In the end, a lame-duck session of Congress in December 1982 cleared the Nuclear Waste Policy Act, a sweeping and complex piece of legislation purporting to put into place a national plan for high-level-waste disposal. The bill set a timetable for establishing a permanent, underground repository for high-level waste by the mid-1990s and provided for some temporary federal storage of waste, including spent fuel, in the interim. In a concession to the emerging power of state governments, the new law gave a state the right to veto a federal decision to place a repository within its borders. Under an amendment by Senator William Proxmire, D-Wisconsin, that helped clear the way for Senate passage, the veto would stand unless both houses of Congress voted to override it. The entire program would be funded through what amounted to a consumer tax on nuclear-generated electricity.[8]

But the difficulty of overcoming the political obstacles to storing high-level waste resulted in a flawed and contradictory piece of legislation. Unlike the earlier legislation involving WIPP, in which House Armed Services chairman Melvin Price single-handedly dictated the terms of what would be done, numerous lawmakers were eager to weigh in on the high-level issue. In the process, they created enough complexities in the name of protecting their constituents to blunt the law's impact. Journalists Donald

Barlett and James Steele described the act as "a political masterpiece of special-interest legislation. . . . It offered a little something for everyone and a catalog of excuses for not doing what government had promised since the 1960s—building a repository."[9]

One result of the legislation's shortcomings was that WIPP would continue to be discussed for years as a potential high-level-waste candidate, much to the unease of New Mexico officials. When Toney Anaya became the state's governor in 1983 after serving as attorney general, he began seeking an ironclad commitment that the plant would not be used for such a purpose. Anaya asked to delay starting construction until he received stronger assurances that the scope of the project would not be changed to include high-level storage. "Our biggest fear is that this may be the camel's nose under the tent," said Sally Rodgers, an Anaya aide.[10] To the same end, Anaya wanted the site subject to NRC licensing. Although that remained objectionable to Price, it made sense to New Mexicans, who saw it as a way to maintain consistency between the forthcoming national high-level site and WIPP.

Energy secretary Donald Hodel refused to accede to Anaya's demands. Instead Hodel assured the governor that the New Mexico site would only be used for limited experiments with high-level waste that would eventually be removed. "I want to re-emphasize that the WIPP facility is not being designed for the permanent disposal of high-level waste, nor has the site itself been characterized for such permanent disposal," Hodel wrote to Anaya in February 1984.[11] But Anaya remained suspicious, largely because he felt agency officials did little to engender trust. "They were a kingdom unto their own," he recalled. "They were very arrogant, and they just didn't want to give us information."[12]

Proposals to use WIPP for high-level storage intensified with the November 1986 election of Republican Garrey Carruthers as Anaya's successor. A former Interior Department official in the Reagan administration, Carruthers was a conservative with close ties to the state's business community. The Energy Department's lack of candor dismayed him, but WIPP fit with his agenda of diversifying the state's ailing economy. As if to underscore that point, Carruthers named as his economic development secretary the combative John Dendahl, who had continued his fervent newspaper-letter-writing campaign in defense of the plant. Dendahl became part of an inner circle advising Carruthers on nuclear issues.[13]

Six months before Carruthers's election, the department had ful-filled its requirements under the Nuclear Waste Policy Act and selected three states for potential high-level-waste repository sites. One was a thick deposit of salt under the flat farmlands of Deaf Smith County in Texas and another was a basalt formation under the Hanford Nuclear Reservation in south-central Washington. The third prospective site was a desert ridge in Nevada, 100 miles northwest of Las Vegas, known as Yucca Mountain. Predictably, the decision met with strong resistance from all three. Under pressure from nuclear utilities, the influential and staunchly pro-nuclear chairman of the Senate Energy Committee, Louisiana Democrat J. Bennett Johnston, joined with Republican James McClure of Idaho in proposing to give states interested in storage as much as $100 million a year in incentive payments.[14]

Momentum for the idea in New Mexico picked up even more steam in September 1987, when an anonymous "white paper" surfaced in the state, presenting a creative idea for resolving the waste storage impasse. A former Sandia National Laboratories scientist, Robert Jefferson, subsequently claimed authorship, along with two others. The paper proposed that New Mexico volunteer to accept the high-level-waste site, which would be situated near WIPP. In return, the state would be awarded the Superconducting Supercollider, an enormous $4.4 billion scientific research facility to be used to smash subatomic particles and probe mysteries of matter and energy. Some two dozen states, including New Mexico, were pursuing the Supercollider, which was expected to generate thousands of jobs. If New Mexico got both projects, the paper claimed, the total economic benefits for the state would reach $62 billion. The idea intrigued Carlsbad's nuclear boosters, and it did not take them long to lobby for consideration for high-level storage. Bob Forrest, a relentlessly cheerful and politically dogged tire store owner who had defeated Walter Gerrells in 1986 to be-come the city's new mayor, told a reporter: "At least, we'd like DOE to come down here and spend a million dollars looking at the area."[15]

The economics of high-level storage soon entranced Carruthers; he saw it as in keeping with New Mexico's long nuclear legacy. "My point was that we had put together the first bomb, we take care of a lot of bombs out here, and every aspect of the nuclear industry except maybe enrichment had been in our state," he recalled in 1999. "And we were not part of the problem but the solution [with WIPP]. Why wouldn't we expect to do this?"[16] But environmentalists loudly protested and, before long, so did New Mexico's congressional delegation. Having endured the bruising political battles seven years earlier over high-level waste, the lawmakers were not eager to repeat them. With WIPP under construction, they argued that they were doing their part to address the broader waste storage problem. First-term Democratic senator Jeff Bingaman also contended that the Energy Department siting process was too far along for New Mexico to appear on the scene. After an unpublicized trip to Washington, Carruthers discovered the reaction from delegation members "was not just no, but hell no. So I dropped it."[17]

Having failed to enlist New Mexico, Senator Johnston engineered the most politically explosive move in the history of U.S. nuclear waste policy. In December 1987, House and Senate negotiators agreed on a plan he devised to take Washington and Texas out of the running for a high-level nuclear waste repository, leaving Nevada's Yucca Mountain as the only site to be studied. The lawmakers also canceled language in the 1982 law that called for a second repository in the more highly populated East. Johnston's thinking was to save money by characterizing only one site in the West instead of the three required. His premise was that the scientific problem of finding a geologically suitable site would be much easier than the political problem of finding a state willing to take the waste. His actions helped to cement suspicions among WIPP's critics in New Mexico that politicians in Washington would, if necessary, act in decidedly undemocratic fashion to achieve what they wanted, especially when it came to states with relatively little political clout. As the Southwest Research and Information Center's Don Hancock noted, "Congress unilaterally chose New Mexico in 1979, and it unilaterally chose Nevada in 1987."[18]

Nevada and New Mexico shared some similarities. Both were geographically vast yet sparsely populated places with a dominant federal government presence, especially when it came to nuclear weapons. At the Nevada Test Site, explosions to assess the reliability of U.S. warheads had been conducted for years. The crucial difference between Nevada and New Mexico, however, was that the latter had established grassroots support as well as at least tacit state government acceptance for burying radioactive materials, allowing WIPP to proceed at an incremental pace. The fact that construction had begun in New Mexico gave supporters a political momentum that was absent in Nevada, which had the gambling industry to provide it with a steady and lucrative revenue stream.

Consequently, Nevada officials struck an unyielding adversarial pose that became the common thread in the state's political fabric. "You cannot get elected here unless you are strongly in opposition to Yucca Mountain," said Robert Loux, executive director of the state's Nuclear Waste Project Office, an agency set up to provide oversight in a fashion somewhat similar to the Environmental Evaluation Group (EEG). Former governor Robert List chalked up such an attitude to the suspicion that arose after the AEC assured the public no harm had resulted from nuclear tests at the Nevada Test Site during the 1950s. "The AEC misled the public, and [now] people don't trust the authorities," List said.[19]

Nevada politicians also echoed the earlier resentment of New Mexico politicians at the idea of storing commercial wastes that the state itself did not generate. Unlike the Carlsbad crowd, which saw itself as lending a helping hand to Uncle Sam while making a little money in the process, Nevadans viewed themselves as the equivalent of a scrawny schoolboy being picked on by a bully. "When you rub all the fog off this window," Nevada Democratic senator Harry Reid said of Johnston's maneuvering, "you look in and you see base power politics at its worst." Reid called the legislation the "Screw Nevada Bill."[20]

Transportation, Hydrologic Issues Surface

While Nevada stayed hostile to the idea of hosting a nuclear waste repository, New Mexico politicians remained willing to negotiate on their concerns. As a result of the agreement Bingaman had reached in 1981 when he was attorney general, New Mexico and the Energy Department formalized several other agreements. In December 1982, the two sides reached a deal that committed the department to seeking money from Congress for upgrading selected state highways for transportation routes to WIPP. It also clarified that the federal government would be liable for any

WIPP-related accidents at or en route to the site. But George Goldstein, chairman of New Mexico's Radioactive Waste Task Force, contended "the real clincher" was language allowing for independent monitoring of transportation. State officials would receive advance notification of what routes would be taken, designate what routes could be used, and travel to out-of-state sites where the waste was produced and monitor its packaging and shipment.[21]

The issue of transportation safety emerged as a central concern. When New Mexico residents learned that railroads would not be used and that truck shipments planned to crisscross many areas of New Mexico—including cities where the state's population was most heavily concentrated—their concerns over accidents heightened. One Carlsbad resident testified at a May 1983 hearing that she would move her family to Oregon because of fears about an accident involving a release of radiation.[22] Transportation also became a fiscal worry for New Mexico's congressional delegation. Although the department promised to seek money to upgrade WIPP roads to carry waste trucks, it limited itself to a pledge to support the state and congressional delegation when they sought a special appropriation from Congress for that purpose.

The resulting chase for money sometimes put New Mexico Republican senator Pete Domenici in a difficult position. As the powerful chairman of the Senate Budget Committee, he was charged with ensuring that overall spending be kept on a tight leash. But in October 1985, Domenici tried to add $16 million for WIPP roads to a transportation bill—a measure he had complained was too large. Although he won the funds, his contradictory stance irritated colleagues. "Now we have $16 million for a little road project in New Mexico, and we are going to have this big, across-the-board cut again because we are not meeting the demands of the Budget Committee," observed Republican senator Lowell P. Weicker Jr. of Connecticut.[23]

Other New Mexicans' transportation concerns centered on the container to be used to truck 55-gallon steel drums of waste to WIPP. Sandia scientists had begun designing a package known as the Transuranic Packaging Container (TRUPACT) in 1978 for railroad shipments but redesigned it for truck use when the Energy Department expressed its preference for that option. The scientists soon hit a technical snag: nonradioactive hydrogen gas was expected to build up inside the container, creating the potential for an explosion. To deal with the problem, a vent was incorporated into the design to, in effect, "burp" the gas. New Mexico officials strongly objected; EEG director Robert Neill argued strenuously in favor of "double containment" to prevent a radiation release, a feature he noted that the NRC required in its waste casks. The Energy Department subsequently rejected the design, prompted again in part by congressional objections. Members of the Senate's Surface Transportation Subcommittee were angry that the department had the legal authority to certify the container on its own, even though it had promised to meet both NRC and Department of Transportation safety standards. "TRUPACT is a single theme that runs through this hearing, and shows that the system doesn't work and there should be an independent group that certifies DOE's regulations," declared Democratic senator Brock Adams of Washington.[24]

Chagrined Energy officials decided to allow the NRC to license the container and awarded a contract to completely redesign it. A prototype of the new 10-foot-high, double-walled model, TRUPACT II, passed several structural tests in 1988, although its seals failed after it was subjected to a 30-minute fire. After further tinkering, the package passed all required tests, including the fire test, a 30-foot drop onto a concrete pad, and a drop onto a metal spike. The TRUPACT II was finally licensed in August 1989. But environmentalists remained unsatisfied, contending that the package needed to be put through a "crush test" to determine what might happen when an oncoming truck or train hit a truck carrying radioactive cargo. They maintained that this would provide the surest test of TRUPACT's effectiveness, though Sandia and Energy officials disagreed.[25]

Once the Energy Department and Sandia dealt with the EEG's concerns over TRUPACT, the state watchdog agency remained satisfied with the tests that the container was put through. On other technical fronts, however, the evaluation group continued to raise concerns about WIPP's safety. In 1985 and 1986, it released studies suggesting that the risk of future contamination of the Pecos River from radioactive waste seepage was greater than the Energy Department had presumed. The Pecos flowed eastward through Carlsbad into Texas, and the potential for any release of radiation into the river would create an interstate controversy with significant political consequences, not to mention health ones.

The EEG focused on the Rustler formation, the underground geologic zone above the 2,150-foot repository. In one study, it said that groundwater in the formation might be replaced or "recharged" by water moving vertically through the formation, increasing the probability that future drilling for oil or gas in the area could lead to radioactive water eventually reaching the Pecos. The EEG's report contradicted Sandia's findings that water 1,500 feet above WIPP's underground chambers had not been replaced by rainfall in thousands of years. The Energy Department quickly criticized the study as "too limited" to supplant its own research. It said that the state group's work "apparently addresses only one facet of a complex geologic and hydrologic system" and that it remained convinced that groundwater recharge was limited in quantity. Such scientific debates had no clear-cut answers, only dueling opinions. They helped perpetuate a climate of uncertainty.[26]

The widespread publicity given to such flare-ups proved to be of increasing irritation to Carlsbad's WIPP supporters, who feared an accumulation of skeptical data and negative news coverage could jeopardize a project that by then had produced more than 600 jobs. In December 1987, a similarly peeved Carruthers struck back. The governor ordered a freeze on the salaries of the five Santa Fe–based EEG scientists, contending they were out of proportion with what ordinary state workers made. He also ordered them to relocate to Carlsbad, putting them closer to the project and, he said, better able to work with Sandia and Energy officials stationed there. Finally, Carruthers demanded that all evaluation group reports needed state approval before publication. "I tried to send them to Carlsbad so they could actually get a view of what they were talking about," he recalled. "They were drawing big checks and pontificating and dragging out the process."[27]

Incensed that their independence was being undermined, Neill and the group's other scientists submitted their resignations. But Bingaman and Domenici, aware of the value of having the group left alone from political meddling, devised a solution. They introduced a bill in March 1988 to transfer the group from under the auspices of the state Health and Environment Department directly to the New Mexico Institute of Mining and Technology, a research university. Carruthers realized he was being overruled and promised to make the transfer administratively, prompting Neill and several of the other scientists to rescind their resignations. The group's headquarters was subsequently relocated from Santa Fe to Albuquerque.[28]

"Proving" Safety with the EPA

For all of Carruthers's concerns about the evaluation group, it remained an organization without any regulatory authority. The question of whether and how a non-Energy Department agency—one with actual enforcement power—should independently "prove" WIPP's apparent safety would become the single most contentious debate over the project.

With the NRC legally kept out of the WIPP licensing process, state officials and environmentalists had long opposed the notion that the Energy Department be allowed to "self-regulate" and declare the plant would accept waste when the department judged it ready. They demanded that the task be given to the Environmental Protection Agency (EPA), an organization in which it placed more trust. The 1982 Nuclear Waste Policy Act directed the EPA to set standards for the management and disposal of spent fuel, high-level waste, and transuranic waste buried in repositories such as WIPP. The following year, state and Energy officials reached a formal agreement requiring the department to comply with "all applicable state, federal and local standards, regulations and laws," including any standards the EPA set forth.[29]

In August 1985, the agency issued its standards, which provided a regulatory framework for assessing disposal safety. The standards established limits on human exposure to radiation and on off-site releases of radioactivity against which to measure performance of individual waste disposal systems. The standards were broken into two segments. The first, subpart A, limited human exposure to radiation from the management, storage, and preparation of waste before its disposal. Subpart B set radiation limits after the wastes were buried. Under the latter, limits were set for individual exposure to radiation from all sources, including drinking water, for 1,000 years after disposal. Overall, though, the standards required that a waste disposal system be designed to handle projected releases of radioactivity for 10 times that period—10,000 years.

Environmentalists and state governments claimed the standards remained too lenient and before long prevailed in court. In July 1987, the U.S. Court of Appeals in Boston threw out the subpart B regulations as a result of a legal challenge from environmental groups and some state governments concerned about being targeted for a high-level repository. The court ruled that the EPA had failed to adequately consider requirements of the federal Safe Drinking Water Act. It directed the EPA to either reconcile the differences between the drinking water standards and subpart B of its repository standards or explain why they were different.[30]

The legal action did not prevent state officials from negotiating another agreement the following month with the Energy Department. That agreement set forth requirements for the agency to have the NRC certify the TRUPACT containers and take steps to prohibit mining above the WIPP site. The agreement also called for the department to in essence pretend the 1985 EPA standards remained in place as it continued its planning. Attorney General Hal Stratton, a Republican, and the Energy Department concurred that the absence of an actual standard posed no serious problem.[31]

But even if the EPA standards were not considered a problem for the state, they would be a major obstacle at the federal level. Even before the July 1987 legal decision to throw out the standards, the Energy Department had wanted as much flexibility over sending waste to WIPP as possible. As a result, it maintained WIPP was a "research and development facility" and would not technically be considered a "disposal facility" until it could be demonstrated to safely hold waste. To that end, the department planned to make about 6,000 shipments until the subpart B standard was met, filling the plant to 15 percent of its total capacity. Energy officials cited having the EPA's assent. Ray Romatowski, manager of the department's Albuquerque office, said EPA officials had informed his agency that radioactive materials were considered "in storage" just as long as they could safely be retrieved.[32]

The EEG and environmentalists, however, considered such tactics a subterfuge to get waste in the ground. To them, 6,000 shipments represented far too much waste for "research and development" purposes. They contended the Energy Department was continuing a pattern established in 1978, when the state was promised veto power over WIPP, by placing political expediency over safety. To Hancock, the site's geologic problems made it clear that "compliance with EPA standards is not just an academic issue. . . . At a minimum, compliance with the standards may require engineered barriers, repackaging the wastes or other measures that would substantially increase the costs of WIPP and could pose significant health risks to workers."[33] Inadvertently or deliberately, the department had put itself in a position to be accused of bending the rules.

As with other contentious decisions, the matter was left to Congress. One of the politicians on Capitol Hill who followed the EPA standards debate closely was Bill Richardson, New Mexico's lone Democratic House member. A former Senate aide, Richardson entered politics in 1978, when he moved to Santa Fe to become executive director of the Democratic State Committee. Within months, he planned a 1980 congressional campaign against Republican Manuel Lujan Jr. He narrowly lost but became a star in his state party. When a new congressional district was carved out of northern New Mexico the next year, he campaigned from dawn to dusk and won. Brash, outgoing, and ambitious, Richardson wasted no time in making a name for himself. "You've got to be aggressive," he told a reporter in his first term.[34] He assumed a prominent role in Democratic Party affairs, particularly as a national spokesman for fellow Hispanics. His gregariousness won him plenty of friends in Congress, but his style rankled others; one Albuquerque newspaper called Richardson the "bad boy" of the state's delegation.[35]

Shortly before his first election in 1980, Richardson began meeting with Santa Fe environmentalists to discuss WIPP. "At the time, I sensed that any type of oversight of health and safety was seriously lacking," he recalled.[36] Richardson's willingness to put up barriers to the Energy Department's plans distanced him from his New Mexico colleagues. Domenici was avowedly in favor of the project, as were Republican House members Lujan and Joe Skeen. The pragmatic Bingaman was less enthusiastic but willing to let it proceed as long as it could independently be shown to be safe. All in all, Skeen—whose southern New Mexico district included the plant—was WIPP's strongest backer. A sheep rancher and veteran of a

Figure 16: *New Mexico representative Bill Richardson (left) and Sandia National Laboratories' Wendell Weart (Department of Energy photo).*

decade in the New Mexico Senate, he was one of the few members of Congress ever elected as a write-in candidate, having defeated former governor Bruce King's nephew David in 1980. Hard-nosed and blunt, Skeen was in the mold of western conservatives who preached a hands-off approach to government regulation of natural resources. He was fond of attacking environmentalists as "self-appointed saviors" who "just don't understand" the problems facing those making a living from public lands.[37]

In May 1987, Skeen and Bingaman, joined by Lujan and Domenici, introduced legislation in the House and Senate proposing to transfer the 10,240 acres of federal and state land at the WIPP site from the Interior Department to the Energy Department—the bureaucratic procedure known as "land withdrawal." Their bill required the Interior Department to pay more than $50 million to New Mexico as compensation for royalties and taxes expected to be lost due to the ban on potash mining and natural gas development on WIPP lands. The jurisdiction over the measure in the House and Senate introduced a new element of complexity into the political debate. Unlike earlier legislation involving the project, which the House and Senate Armed Services committees exclusively controlled, the bill was referred to three House committees: Interior, Energy and Commerce, and Armed Services. It also was the province of the Senate Energy Committee in addition to the Armed Services Committee. The multiple committee referrals made it impossible that a single politician could impose his will on the project, as Price had done in 1981. But it also opened the legislation up to increased scrutiny from panels with vastly differing priorities—and thus added to the number of hurdles that WIPP had to clear.[38]

Tempers Flare over Land Withdrawal

As hearings in Congress began on the bill in the fall of 1987, it became apparent that the legislation would not glide to passage. Richardson chose not to cosponsor the bill, citing the lack of a commitment to obtaining $190 million for highway bypasses around Santa Fe and other cities on the route to WIPP. Instead of trying to authorize funding in the land withdrawal bill, the other delegation members decided to let New Mexico pursue efforts to get the money added to the six-year-old agreement that it had signed with the Energy Department. "If we're going to get into building roads, we're going to get mired down," Skeen said.[39] The lack of consensus among the delegation sent a mixed signal to other legislators. On most parochial matters, members of Congress generally defer to the wishes of colleagues in the states that legislation directly affects.

Other problems arose that affected WIPP's credibility. One of the main issues was the seepage of brine into the underground repository—an issue made public by critics. University of New Mexico geologist Roger Anderson caused a stir in January 1988 with his contention that salt-saturated brine water was entering the chamber through shafts drilled from the surface and through the fissures in the walls at a rate he said could make WIPP unfit for permanent storage. Anderson and other scientists contended that the moisture of the brine could corrode the steel drums, making it nearly impossible to retrieve the waste. Another concern was that after the underground chamber was sealed, radioactive water could reach the surface through cracks in geologic formations or through wells drilled in the future. "There are fundamental geologic and hydrologic problems," said Lawrence Barrows, a former Sandia geophysicist who joined Anderson in criticizing WIPP's suitability.[40] To Anderson, the discovery of brine seepage "refuted the original assumption that the [salt] beds are dry."[41]

Sandia scientists, along with Energy officials, dismissed the criticisms as overstated and inaccurate. They said they conducted tests to measure the flow of brine into the chamber and its effect on the waste drums, with the results showing not enough would dissolve the salt. "Within 70 years, long after we are through down there, the mine will look like a perfectly solid rock," Sandia's Wendell Weart told *The New York Times*. "We do not foresee that this site will turn into a fluid regime." But EEG scientists were not as certain. They estimated that much more water could leak into each chamber, rendering WIPP unusable.[42] It was another case of dueling science that made controversy flourish and consensus difficult.

Domenici tried to settle the matter by calling in the National Academy of Sciences. The academy's WIPP review panel concluded that seepage of water into WIPP "probably isn't a serious problem." Panel chairman Konrad Krauskopf said members were "unconvinced" that a buildup of radioactive water could occur underground. But the panel also recommended that only the amount of waste needed for five years of research be placed underground. The result was a scaling back of the proposed initial waste shipments from 125,000 drums to 25,000, an amount equal to 3 percent of WIPP's total capacity. That total was still enough to fill four football-field-size storage rooms.[43] Krauskopf, however, saw the amount of waste as trivial compared with what scientific data might be gleaned. "I think we'd learn a good deal if we started putting it in," he said. "Chances are very good that it will just stay there."[44]

Such arguments may have seemed sound to scientists, but they made little political sense in the climate of uncertainty that had developed. The brine issue fueled the resolve of environmentalists—and, by association, Richardson—that subpart B of the EPA's standards had to be met. At Richardson's urging, the House Interior Committee approved a land withdrawal bill in July 1988 mandating that WIPP comply with both sections of the standards. He included an amendment giving the EPA two years to comply with a court mandate to rewrite subpart B or explain why that section failed to match the Safe Drinking Water Act's stringent groundwater standards. But the Energy and Commerce and Armed Services committees resisted on the standards issue. Armed Services, in particular, did not want the EPA involved in what it still regarded as a defense project.

In the Senate, Domenici and Bingaman continued to seek a middle ground, announcing their intention to modify their proposal. In late September, they succeeded in pushing their bill through the Senate Energy Committee.[45] But such actions were still not enough to get the New Mexico delegation on the same political page. By October, having failed to placate Richardson, the other four delegation members declared the withdrawal bill dead for the year. With so little time remaining in the legislative session, Bingaman noted, "The only way to pass a bill at this stage is for the delegation to present a united front." Domenici and Skeen angrily accused their Democratic House colleague of failing to negotiate in good faith. "It's obvious to me that unless it's all his way, you aren't going to get a bill," said Domenici. Richardson, for his part, held fast to his belief that both parts of the EPA standard needed to be met. "It struck me at the time that the Carlsbad community, the governor, the delegation were trying to ram WIPP down everybody's throat," he recalled.[46] He sought to blame the Reagan administration's Energy Department and secretary John Herrington. "DOE needs some leadership here in Washington that is knowledgeable about [WIPP]," he said at the time. "I had the impression that with Secretary Herrington, this was not a high priority with him."[47]

Despite the disappointment of WIPP's supporters, it was apparent that the project's failure to open as scheduled could not be pinned entirely on Congress. The facility had not complied with numerous requirements: receiving the NRC's certification for the TRUPACT II container, obtaining a permit from New Mexico to bury "mixed waste" with both hazardous and radioactive characteristics, and finishing a safety analysis report for the state, outlining risks to workers and the public. In a letter to Herrington, Domenici also cited the department's failure to formulate a plan for experiments with the waste that could prove acceptable to the EEG and Academy of Sciences.[48] The senator had warned for months that the department's October goal for opening was unrealistic, but Energy officials stuck to the date, contending they needed to establish a fixed deadline just as a highway engineer does for a freeway. Such a move may well have made bureaucratic sense, but it created little public confidence when it was not met.

The supporters of WIPP, meanwhile, tried to deflect blame by responding that the plant faced an ever growing and constantly shifting number of regulations and expectations from Congress and the public. "The problem is that everything that's required is a moving target—it keeps changing," said Jefferson, the former Sandia scientist. But others conceded they had not dealt with all public concerns. Sandia's George Allen acknowledged, "There is a fairly well-perceived credibility problem that will have to be addressed."[49]

Andrus Takes Center Stage

From Idaho, Andrus watched with disgust as WIPP's prospects unraveled on Capitol Hill. The governor had tried to lobby Richardson, arguing in a letter in July that his state's residents "have been good citizens and have been temporarily storing low-level and transuranic waste until a more permanent repository is established . . . but we have had enough broken promises."[50]

Andrus was aware he lacked the legal authority to prevent waste shipments that ordinarily would have gone to WIPP from continuing to be sent into his state. But he was also aware of the leverage from taking on a federal agency that already had been placed on the defensive, particularly on a subject that had received as much increased public attention as WIPP had. He refused to see the technical concerns as an impediment. "A nation that can send a man to the moon and bring him back safely can find a solution to this problem," he said years later. "It's not a question of scientific ability. It is a question of political will."[51] The debate over nuclear waste storage thus became firmly cast on political, as opposed to technical, grounds.

And so, after returning from his WIPP tour, Andrus announced on October 19 that his state's borders would be closed to further waste shipments from Rocky Flats. "I'm not in the garbage business anymore," he declared.[52] Idahoans overwhelmingly approved of the decision. *The New York Times* carried a photograph of a railroad car that had made it over the border and sat on a siding in Blackfoot, Idaho, waiting to be sent back to Colorado.

The governor had stationed a state trooper near the railcar "for safety purposes, and because I didn't trust the Department of Energy." The image of the trooper appearing to block the train helped bring home the defiance that a state felt toward continued federal impositions. "We provided a visible trigger to a complicated subject," Andrus wrote in his autobiography. "By closing its borders, Idaho brought the whole issue of nuclear waste—and the federal government's poky progress—into the national spotlight. With its fear of a spreading rebellion, the Department of Energy was willing to talk."[53]

Other politicians saw Andrus's actions as pure political theater. "He was certainly quite a showman to say, 'I'm going to take on these dudes and we're going to get some resolution to this, because this is my state,'" Carruthers recalled.[54] Another governor who noticed was Colorado's Roy Romer, a popular Democrat like Andrus. Romer warned that if the waste destined for Idaho continued to pile up at Rocky Flats, it would exceed within three months the storage limit negotiated between Colorado and the Energy Department. That would mean the only facility in the United States manufacturing plutonium triggers for nuclear bombs might have to close. Environmentalists decried the thought of waste accumulating at a facility so close to metropolitan Denver. Like Andrus, the Colorado governor milked as much political mileage as he could out of the situation. USA Today ran a front-page color photograph of an unsmiling Romer with his fist raised and an enlarged quotation: "I'm saying to the federal government, 'Get off your duffs.'"[55] The article was just part of a torrent of news stories that played up the conflict among the governors—a conflict that The Washington Post dubbed "a high-stakes game of plutonium poker."

Behind the political posturing, however, was a growing national movement that had put the Energy Department on the defensive. At its production plants, state officials and environmentalists were raising serious concerns about the agency's ability to adequately protect public health as it operated under the Cold War veil of secrecy. At the Feed Materials Production Center in Fernald, Ohio, information came to light that the agency had known for 20 years that thousands of tons of uranium waste were being secretly released into the environment, exposing thousands of workers and neighboring residents to radiation. Local citizens filed a lawsuit

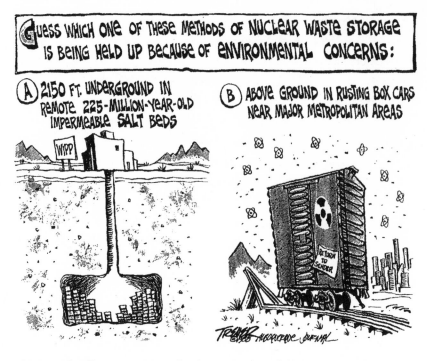

Figure 17: *The* Albuquerque Journal's *John Trever satirized the controversy over delaying WIPP's opening in favor of temporary waste storage in Colorado in a 1988 cartoon (Courtesy of John Trever).*

seeking $300 million in damages. In Washington State, reporters steadily gathered evidence that during the 1940s and 1950s, more than 440 billion gallons of chemical and radioactive liquid waste had been poured into the ground at the Hanford Nuclear Reservation—including enough plutonium to build more than two dozen nuclear weapons. Newspapers also reported that thousands of "downwinders" had been showered for years with radiation through accidents, routine emissions, and deliberate experiments.[56]

The revelations began to alarm and antagonize residents of the cities clustered around the weapons complexes, where people had gratefully accepted the jobs and income that nuclear weapons work provided. At a November public hearing in Twin Falls, Idaho, on a proposed nuclear production reactor, dozens of people spoke of their uneasiness over accepting more risks. "We've been a nuclear dumping ground and national laughingstock for too long," said singer Peter Cetera of the pop-rock group

Chicago, a resident of nearby Ketchum.[57] Similar concerns reverberated across Capitol Hill. Thirty-one of the House Armed Services Committee's 51 members wrote to President Reagan and Herrington to express "grave concern at the seriously eroding public confidence" in the nuclear weapons complex.[58] Congress eventually approved the creation of an independent oversight panel to review safety and environmental conditions at some weapons production facilities.[59]

Facing an accumulation of negative publicity, as well as increasingly hostile state officials, the department called an extraordinary summit meeting with Andrus, Romer, and Carruthers on neutral turf in Salt Lake City. Each governor came to the December meeting armed with a list of demands. In addition to having wastes sent to WIPP instead of his state, Andrus wanted a commitment to dig up, repackage, and remove materials at Idaho National Engineering Laboratory that had been buried in shallow trenches before 1970. Romer also wanted Rocky Flats wastes to go to WIPP, as well as the ability to continue to enforce the state-mandated storage limit at the site. Carruthers wanted the plant open but also sought $200 million for upgrading state highways.

The governors met behind closed doors with Deputy Secretary Joseph Salgado, who outlined the national security implications of a potential shutdown at Rocky Flats. He said the department would continue to support Congress's attempts to pass a WIPP land withdrawal bill instead of attempting such a procedure administratively. Salgado also promised the department would appoint a task force to look for sites around the country where waste could be stored in the event that WIPP could not take the materials as scheduled. And he said the department would update its 1980 environmental impact study of the project, even though such a study would push the earliest expected opening date to August 1989.[60]

All sides emerged from the discussion appearing outwardly hopeful. "Before this day, we were three governors with three different sets of problems," Carruthers told reporters. "We came together with DOE; I think we have a good-faith effort now." Romer added that "I am willing to go home and say, 'I think there's a solution.'" Andrus said he would reconsider allowing waste from Rocky Flats into his state in January, as long as Congress made progress toward passing a land withdrawal bill. "I believe that DOE has put in motion a schedule that can work," he said. "Whether it works or not, I'm not yet prepared to say."[61]

In the months ahead, however, the department's schedule would remain impossible to meet. As a result, the governors would wield their newfound political clout in ways that would continue to remain difficult for the agency to ignore, even under a new presidential administration and new Energy secretary who approached waste cleanup—and WIPP—far more seriously than any of his predecessors.

5

"I Love WIPP!"

1989: AN ADMIRAL AT THE HELM

James D. Watkins described himself as a strategist. "I love strategies and I believe a strategy is critical," he said in 1987.[1] The day before he was confirmed as secretary of Energy in February 1989, he ordered his senior staff to outline his most urgent "near-term" priorities. Among the seven identified, along with a new national energy plan, was securing legal transfer of the 10,240 acres of land at the Waste Isolation Pilot Plant (WIPP).

The "land withdrawal" was a difficult milestone that required placating the safety and environmental concerns of Democratic legislators who had thwarted passage of such a measure a few months earlier. Without the transfer, however, department officials assured Watkins that the nuclear waste experiments it proposed at the New Mexico site could not begin. And Watkins saw the need for quick action. "We're serious about WIPP," he told the Senate Energy Committee at his confirmation hearing. "We have to start the flow of waste."[2]

By showing that radioactive materials could be stored in New Mexico, Watkins believed, the department would be on the way toward gaining the public and political confidence it needed to open Nevada's Yucca Mountain for the much larger task of high-level-waste storage. Opening WIPP would demonstrate that the department could reverse course and take care of one of the nation's thorniest problems. "WIPP was, to me, the ideal place to get public acceptance, because you already had local acceptance," he recalled. "WIPP was not treated as the precursor to Yucca Mountain, and it should have been."[3] The secretary's aggressive efforts would bring the project closer to opening but end up spawning a series of new legal, environmental, and political conflicts—all of which would have far-flung repercussions.

When it came to nuclear issues, Watkins was charged with doing more than just opening the repository—he had to modernize and clean up the entire weapons complex, which was reeling from revelations of environmental and health atrocities. To Senate Energy chairman J. Bennett Johnston of Louisiana, the task represented "perhaps the most daunting challenge that we have ever given to anybody in the Cabinet."[4] But Watkins was familiar with both nuclear energy and tough assignments. A conservative Californian whose grandfather helped found Southern California Edison, he was a veteran of four decades in the U.S. Navy. He retired as an admiral in 1986 after four years as chief of naval operations, the capstone to a glittering military career. He served tours in Korea and Vietnam and had been commanding officer of the USS *Long Beach*, the navy's first nuclear-powered surface ship. Later he became chief of naval personnel, commander of the Sixth Fleet, and commander in chief of the U.S. Pacific Fleet. A trained nuclear engineer, he had been a protégé of Admiral Hyman Rickover, the acerbic and legendary father of the "Nuclear Navy." As he rose through the navy, however, Watkins picked up a reputation for questioning the status quo. In the 1970s, as the women's rights movement took root, he insisted that women be given greater opportunities. His strong moral principles gave him a nickname around the Pentagon: "The Cardinal."[5]

Even in retirement, Watkins asserted himself. At President Ronald Reagan's request in 1987, he was brought in to lead a commission that had been formed to study acquired immunodeficiency syndrome (AIDS), then a new public health worry. The group's credibility was under fire and its members were bickering over ideology when Watkins agreed to become its chairman. He hired new staff and persuaded the divided factions to agree, and within seven months, his panel surprised health experts by issuing a comprehensive set of suggestions regarding health care for AIDS victims. Among its recommendations was that legal protections be given to those who tested positive for human immunodeficiency virus (HIV), the virus that causes AIDS. He picked up another nickname: "Radio Free Watkins."[6]

Watkins's work in resurrecting the AIDS commission impressed Vice

President George Bush, who had broken ranks with the Reagan administration to support its entire package of recommendations. So it hardly came
as a surprise in January 1989 when Bush tapped the 61-year-old Watkins for
Energy secretary. The admiral's nuclear-oriented background stood in sharp
contrast to that of his predecessors—two lawyers and one dentist—and most
of the reaction was positive. *The New York Times* described Watkins as an
"unusual leader" with "forceful opinions and [a] record of independence."
The Washington Post noted his "political skill" and "competence."[7]

To successfully implement his strategies, Watkins felt he needed to overhaul how the department functioned. "I knew it was a bad cleanup mess,
but it was a management mess as well," he said.[8] At his confirmation
hearing, Watkins lamented what he called an "antique, out of date" administrative system and process oriented too heavily toward weapons production
at the expense of environmental concerns. He promised Johnston and
other senators he would personally steer an activist course in reshaping
the agency: "There is an urgent need to effect a significant change in its
deeply embedded 35-year culture."[9]

As he had done with the AIDS commission, Watkins required little time
to make his mark. His imposing bearing and no-nonsense temperament
made him a formidable—and feared—presence. "The Admiral" became
infamous for barking criticism at aides whose performance displeased
him. His desire for control extended into his personal life; shortly after
becoming secretary, he prevented one of his daughters from marrying a
man he considered unsuitable for her.[10] Watkins sought results through a
centralized approach carried over from his military career—a style that cut
through bureaucratic obstacles but invited confrontation with critics.
His hands-on attitude was in the tradition of his mentor, Rickover. "He
was right in your backyard every time you turned around on one of those
nuclear-powered ships," Watkins recalled. "If you thought he was way back
in Washington and you were out there—no. Big Brother was watching
you everywhere, and that's the way it has to be with things nuclear."[11]

In March, Watkins promised Congress he would come up with a
comprehensive five-year plan that would cover nuclear waste management,
cleanup of inactive waste sites, and other corrective actions. The plan's
first priority would be to bring the department's waste management and

environmental activities into compliance with state and federal laws. To develop the plan, Watkins named Leo Duffy his "waste czar," or special assistant for coordinating defense waste management. Gruff and self-confident, Duffy had been an environmental consultant and engineer for several Energy contractors and had served under Rickover. He also knew how to get along with Watkins. "As long as you kept him informed, you were in good shape," Duffy recalled.[12]

Shortly after joining Watkins's staff, Duffy assured an Albuquerque audience in an impassioned speech that the plant would not open unless it was proved safe. But environmentalists in attendance were dismayed when they asked who would determine its safety; Duffy answered it would be Watkins and that WIPP most likely would be judged safe in time to receive waste in September 1989. Such remarks were among the earliest indications to them that the department remained deaf to external input.

Flaws Cited in Revamped Environmental Study

Energy officials, however, were determined to portray their agency as intent on listening to what others had to say. By April, four months after the Salt Lake City meeting between Energy officials and the governors of Colorado, Idaho, and New Mexico, the department made good on its promise to issue a draft supplement to its 1980 environmental impact statement for WIPP. It also announced a series of public hearings on the plan to be held at various cities near department sites on the shipping route to the plant. To indicate its desire to reach a wide audience, it scheduled the sessions on or near waste shipping routes between Pendleton, Oregon, and Atlanta. After environmentalists complained, the department added extra hearings in New Mexico. "We have always maintained that despite all the political pressures and all the hullabaloo, we will not compromise on safety," said Richard Marquez, an Energy official in Albuquerque. "I think that by going around to the states and holding public hearings like we are, we can show people where we're at."[13]

The updated, 1,000-page study was a wide-ranging document. It assessed such issues as the safety of the containers transporting waste to WIPP, the geologic and environmental consequences of an underground disturbance, the risks of radiation exposure, and the question of whether brine seepage

Figure 18: Energy secretary James Watkins (right) with his "waste czar" Leo Duffy during a visit to WIPP. In the background is Colorado governor Roy Romer (Department of Energy photo).

represented a hazard—all questions uppermost in critics' minds. The study's overall conclusion echoed the one of nine years earlier: that the plant could indeed be operated safely under the present methods and timetable. The widespread thinking among the project's management and scientists at Sandia National Laboratories was that little had changed to suspect such a belief merited any reconsideration.

The outside view, though, was markedly different. Some observers questioned the relative swiftness with which the conclusions of the new study were reached, not to mention the conclusions themselves. Among the skeptics was the New Mexico Environmental Evaluation Group (EEG), which issued a review charging that the environmental plan "does not provide adequate justification" to support shipping waste to WIPP. "The document contains mistakes in calculations, reflects an erroneous knowledge of the history of the project, presents tables without units, and displays an indifference to the statistical precision of predictions," the group said. It pointed to the failure to address the potential effects of an underground fire and the scant attention paid to the environmental consequences of a scenario in which waste would be kept out of WIPP and stored elsewhere. "It appears that the draft [study] was written with a predetermined conclusion to accept the proposed plan," their report said, "and that alternatives were not seriously considered."[14]

The group's comments were part of a series of harsh statements it made about WIPP's scientific direction that would get wide attention in 1989. In May, its deputy director, Lokesh Chaturvedi, reported sizable cracks in the ceilings and floors of two large waste storage rooms at WIPP. The cracks, according to Chaturvedi, served as evidence that the repository's salt walls were closing in faster than expected. Although the group had known for several years that the pace of underground "salt creep" was increasing, the presence of extensive fractures in the ceilings of two rooms came as a surprise—as did the department's decision to use rock bolts to stabilize the ceilings of seven newer storage rooms.

Chaturvedi's concerns were given one of the nation's most prominent media forums. *The New York Times*'s Washington environmental correspondent, Keith Schneider, wrote an article based on the scientist's comments under the headline "Nuclear Waste Dump Faces Another Potential Problem." Brash and idealistic, Schneider had already given fits to project officials in New Mexico. After visiting the site in 1988 and writing a lengthy story about the debate over its scientific flaws, Schneider began aggressively pursuing environmental hazards at other Energy sites. His articles, which often received front-page play, helped pull the problems of the nuclear weapons complex onto the national agenda. At WIPP, Energy Department officials insisted to Schneider and other reporters that the cracks were expected and, in fact, welcomed because they enabled scientists to learn more about salt closure. "We don't think this is a problem," said James Bickel, assistant manager for projects and energy programs at the Albuquerque office. "But we live in a real world, and a world of perceptions. If people think this is a valid concern, then we need to respond to it." Chaturvedi, however, remained perturbed that the department had not informed his group. "They should have told us about it," he said. "We

Figure 19: *Environmental Evaluation Group deputy director Lokesh Chaturvedi (courtesy Lokesh Chaturvedi).*

are here to know everything about the site."[15]

The incident was not the first time Chaturvedi's forthrightness would anger the Energy Department, nor would it be the last. A native of India, he came to the United States in 1963 to earn a master's degree in civil engineering from Purdue University and a Ph.D. in geology at Cornell University. He joined the evaluation group in 1982 after teaching geology at New Mexico State University and several other universities and becoming known for his work on geothermal energy. At the time of the move to study WIPP, he figured it would make an interesting temporary stopover before resuming his academic career. "Everyone thought it would be over in a couple of years and that the waste would come in '86 or '87," he recalled years later, chuckling. "No one ever thought this would become a lifetime job."[16]

Although Chaturvedi endorsed the concept of WIPP, he was quick to outline his problems with it. By 1989, the department's focus on shipping waste for testing at the expense of determining its suitability for permanent storage deeply bothered him. The Energy Department had first proposed the tests to measure rates of gas generation from waste drums in WIPP underground rooms in October 1987, with the idea they would help prove to the Environmental Protection Agency (EPA) how the repository would behave over the long term. The department, however, did not provide the first draft of a plan for the tests until the following March. Because of what Chaturvedi called "a large number of deficiencies in the plan and the lack of connection between the goals and the design," the plan was abandoned. In September 1989, Chaturvedi wrote a report that looked at preliminary drafts of the types of tests under consideration. He remained skeptical: "It does not appear likely that the experiments will yield a rate of gas generation low enough to be acceptable for prediction of [the] satisfactory long-term performance of the repository. . . . The results will most likely not help in showing compliance with the EPA standards."[17]

The department, however, continued to insist on the tests. The situation prompted Chaturvedi to remark that the agency should be allowed to send a single shipment to Carlsbad so officials could boast they had "opened" WIPP before turning to the business of ensuring its long-term suitability. Claims of needing waste for testing were "mere excuses to put some waste in the ground," he said. "Successive DOE managers hoped that once some waste was put in the ground, the ice would be broken, and as a *fait accompli* the public would more easily accept emplacement of a million drums."[18] Chaturvedi's candor made him popular with reporters and critics while earning him scorn from supporters. "I think Lokesh was always against the project," Duffy said.[19] But Chaturvedi went to considerable lengths not to be viewed as a hostile zealot. "I'm not anti-WIPP," he explained. "I'm pro-taxpayer."[20]

"Speak Now or Forever Hold America's Waste"

Chaturvedi would, in time, become an even more influential contributor to the climate of conflicting science in which WIPP now remained rooted. But in terms of influencing the public perception of the project, his opinions would not approach the impact of the hearings on the environmental study held in New Mexico during early June. It was at those sessions that the long-building opposition to WIPP was galvanized and given both a prominent public and media platform. Such a situation arose, in part, because critics of the project made a concerted push to fill the vacuum created by the Energy Department's inability to harness public consensus on national nuclear issues.

Several relatively powerful environmental groups already had begun to fill that vacuum by the late 1980s. The involvement of the groups helped foster the notion that WIPP was not just a state issue but a national one, with serious environmental implications. Unlike the loosely organized New Mexico groups of the past that held placard-waving demonstrations, the national groups relied on the courtroom to directly address the department's accountability. Two of the organizations showing a particular interest in WIPP were the Natural Resources Defense Council (NRDC) and the Environmental Defense Fund (EDF). Both groups had proved a formidable nuisance to the federal government and corporate world. In February 1989, the television show "60 Minutes" aired a report based on an NRDC study that said a chemical growth regulator used on apples, Alar, was "the most potent cancer-causing agent in our food supply." The story set off a panic, with supermarkets dumping apples and apple products. Farmers, politicians, and others angrily accused the group of overstating the chemical's effects and using science to provoke hysteria. But the group countered that its actions had helped lead the EPA to ban Alar as a public health risk after finding it to be a potential carcinogen.[21]

Both organizations employed young, aggressive lawyers who were skeptical of the Energy Department's promises. The NRDC's attorney, Dan Reicher, was a former Massachusetts assistant attorney general who had worked on a presidential commission to study the Three Mile Island accident. Under his watch, the group established the legal precedent for holding the Energy Department accountable for hazardous waste at its nuclear sites.[22] The EDF's attorney, Melinda Kassen, served as a member of the Rocky Flats Environmental Monitoring Council, one of the public watchdog bodies that monitored conditions at the Colorado weapons plant. She became a widely quoted authority on both Rocky Flats and WIPP. In December 1987, she warned the department it could not open the New Mexico site without first revising its environmental impact statement—something the department subsequently agreed to do at its meeting with governors in Salt Lake City a year later.[23]

In March, Kassen and Reicher issued another warning on WIPP. Their groups formally put the department on notice that they would file a lawsuit unless the department addressed their concerns. Among their demands was that the EPA issue new radioactive waste disposal standards before WIPP opened, a process expected to take two more years. Although the groups had made similar demands in the past, their arguments received attention because their cause had a new ally: the state of Texas. Officials in Attorney General Jim Mattox's office noticed the EEG's concerns about the possibility of radioactivity reaching the Pecos River, which flowed into their state. They also worried about transportation, since shipments from half a dozen different plants would travel through Texas on their way to WIPP. The involvement of a state government helped lend credence to environmentalists' arguments that antinuclear extremism dictated their opposition to WIPP. "We just want [the department] to follow the law," said Texas assistant attorney general Renea Hicks.[24]

In New Mexico, meanwhile, grassroots organizing efforts involving WIPP continued. At the forefront stood Concerned Citizens for Nuclear Safety (CCNS), a group that a handful of Santa Fe residents had founded a year earlier. The group objected to the department's plans to ship waste from Los Alamos National Laboratory through Santa Fe on St. Francis Drive, a busy four-lane commercial street west of the downtown business district. Business and restaurant owners joined Concerned Citizens to contend that Santa Fe's mountain beauty and its historic uniqueness were prized and vulnerable assets. Clothing store owner Richard Johnson, one of Concerned Citizens' founders, summarized the group's central concern: "Just one accident will destroy the economic base of this town. We have nothing to gain and everything to lose."[25]

Energy Department officials routinely answered such comments by noting that trucks carrying highly toxic substances such as propane and gasoline already were regular and safe users of the road. Besides, the officials pointed out, Santa Fe would see relatively few WIPP shipments, and those that would pass through town were expected to employ an as yet unbuilt highway bypass. But such arguments failed to take into account the fact that the public perceived radioactive materials as posing especially unique threats. To that end, Concerned Citizens' main argument that an accident involving

Figure 20: A 10-year old
protestor at a 1988
anti-WIPP rally in
Santa Fe (Santa Fe New
Mexican photo).

a shipment could scare off would-be visitors and demolish the city's
tourist-dependent economy managed to unite diverse segments of the
city's multiethnic and multieconomic population. The threatened loss of
income cut across all ideological beliefs and biases. Concerned Citizens
urged merchants to place "Another Business Against WIPP" signs in their
storefronts; dozens complied. It ran a series of radio spots asking, "Citizens
of New Mexico, can we trust the Department of Energy?" It lobbied local
city officials, persuading the city council, school board, and Board of Realtors
to adopt resolutions of opposition. Such tactics had a measurable impact. In
the fall of 1988, Santa Fe's government commissioned a telephone poll of
nearly 400 adult residents. When asked to name the major issue facing the
city, nearly one in five respondents cited WIPP, a facility more than 250
miles away.[26]

Concerned Citizens members portrayed the hearings in New Mexico as an obligatory exercise for anyone troubled with the government's nuclear waste policy. In doing so, they again helped assign WIPP a larger significance in the overall environmental cleanup debate. Its members made clear that the intended audience for the hearings would not be department officials as much as members of Congress who would be considering the land withdrawal bill. Emphasizing the public's duty to air its views, they erected billboards around the state proclaiming, "Speak Now or Forever Hold America's Nuclear Waste." In the end, more than 600 people signed up to testify at hearings in Albuquerque and Santa Fe, compared to the 100 or so who had done so at earlier sessions in the other states.

The New Mexico hearings were a watershed for the anti-WIPP movement. The vitriolic sessions proved to be a public relations embarrassment for the department, drawing extensive regional as well as national news media coverage. Over four days of sessions in the two cities, the speakers emphatically, and often flamboyantly, registered their dismay. Because the hearings were structured in such a way that department officials could only listen to testimony after making a short opening presentation about the findings of their study, few if any of the opponents' emotional assertions were disputed. Outside the Albuquerque hearing, a truck carrying workers dressed in radiation protection clothing staged a mock nuclear accident. Amid a carnival-like atmosphere of street performers and musicians in Santa Fe, participants cheered the scathing criticisms of the Energy Department. Comparisons were made to the totalitarian People's Republic of China and Star Wars movie villain Darth Vader.

The testimony drew from a cross section of New Mexico's multiethnic population. Even former governor Bruce King dropped by to offer a few criticisms. One of the Native Americans who spoke, Patsy Jojola of Isleta Pueblo, drew prolonged applause when she declared, "Let it be known that to transport this waste through our land is disrespectful, inhuman and interfering with the laws of nature." Former Southwest Research activist Charles Hyder, who had drawn international attention for fasting 218 days in front of the White House in support of nuclear disarmament, blasted the study as "a betrayal of the U.S. people, a perversion of the English language [and] a corruption of public funds."[27] Others acted out skits or performed songs; one man strummed a guitar and sang, "What if brine begins to seep, and it creeps up from the deep? As we sow, so shall we reap. WIPP it!" It was such a catchy tune that even Sandia's Wendell Weart found himself tapping his foot.[28] Energy officials were careful to be diplomatic. "People are free to speak on their allotted time on anything they want," Marquez said.[29]

The testimony at the hearings indicated that public unease over nuclear energy had intensified since the April 1986 explosion of a reactor at Chernobyl in the Ukraine, the worst nuclear accident in world history. The explosion killed three people, and 28 emergency workers died within three months. United Nations health officials later said the number of those likely to develop serious medical conditions because of delayed reactions to radiation exposure would not be known until 2016.[30] In assessing the department's public perception problems on nuclear weapons cleanup and WIPP, Bickel acknowledged that "Chernobyl had a monumental effect on us, probably more than on the Soviet Union."[31] One bumper sticker seen on cars around Santa Fe bore the warning "Three Mile Island/Chernobyl/WIPP." Such protests ignored the vast differences between the three projects in favor of making explicit the fear of all things nuclear.

To that end, it hardly helped the Energy Department that the company it had hired to operate the plant was Westinghouse Electric Corporation, the world's biggest manufacturer of nuclear reactors. Westinghouse became the top reactor builder largely as a result of its bargain-basement sales strategies in the 1960s, when it offered electric utilities a 20-year supply of uranium fuel for any reactor it built. Twenty-seven utilities agreed, leaving the company obligated to provide more fuel than it had—a problem that led to numerous lawsuits. Before it won a five-year WIPP contract in 1985, Westinghouse had maintained a presence at the site since 1978, when it was a technical contractor. The Pittsburgh-based corporation managed several other Energy sites, including several that had made headlines with revelations of environmental mishaps such as Washington's Hanford Reservation.[32]

During the WIPP hearings, project proponents did their best to communicate their views. They maintained the plant was a necessary new way to help solve the national storage crisis. But their approach was inadequate in addressing the public's deep anger. "There is a lot of emotionalism," said John Arthur, an Albuquerque Energy Department official. "As far as looking for an adequate resolution of that emotionalism . . . well, it's mighty hard to do." A busload of Carlsbad residents tried to help by showing up at the Albuquerque hearings. "This project is the best thing that ever happened to our city in the last 15 years," Mayor Bob Forrest testified. But Forrest and the others were jeered; a few were spat on.[33]

Although the broad issues of government mistrust and the potentially dire consequences of imprecise science were the dominant themes at the hearings, they were not the only ones. Concerned Citizens members sculpted their testimony around the Energy Department's position that it bring waste for tests before meeting the EPA's subpart B regulations governing long-term disposal for up to 10,000 years. Hundreds of protestors waved printed red-and-white placards that pointedly proclaimed, "WIPP Must Meet NEW EPA Standards." Many of the speakers capitalized on the issue. One of them, former New Mexico Health and Environment secretary Joseph Goldberg, asserted that the standards were "the only way to protect the public."[34] The cumulative effect of such testimony was that the technical merits of the environmental study—the ostensible purpose of the hearings—became superseded by the question of whether the long-term EPA standards should be met and whether the department was harming the public by not doing so. Energy Department officials effectively had their agenda swiped from them.

Skepticism Continues in Congress

The debate over scientific uncertainty was not just between the department and citizen activists. On Capitol Hill, a separate campaign was being waged against the agency. Following the failure of the land withdrawal legislation in 1988, lawmakers who had misgivings about the standards issue and WIPP overall sought to build a case against the plant in preparation of Congress's next attempt at a bill.

The leading skeptic was Mike Synar, a Democratic congressman from Oklahoma and persistent critic of governmental waste and abuse. Elected in 1978 just after his twenty-eighth birthday, Synar was a cocky and often

Figure 21: A 1989 anti-WIPP rally in Santa Fe (Santa Fe New Mexican *photo*).

caustic populist. He took on cigarette companies, introducing legislation that would ban tobacco advertising and promotions. He advocated loudly for gun control, drawing the enmity of the National Rifle Association. And his efforts to raise fees on ranchers who grazed cattle on public lands made him a widely vilified figure in much of the rural West. Synar's unswerving confidence in his causes earned him a reputation as a serious if controversial legislator. "Do I make people mad?" he said. "Absolutely. Has it caused me a tremendous amount of problems? Absolutely. And am I on the edge all the time, being in political danger? Absolutely. But that's what you have to do to make a difference."[35]

Synar used his chairmanship of the House Government Operations Committee's panel on environment, energy, and natural resources to conduct aggressive inquisitions into programs he believed were not functioning properly. In doing so, Synar attracted national news media coverage for issues that otherwise might have received less scrutiny. By taking on the Energy Department, Synar earned admiration from environmentalists. "If it hadn't been for Synar, there wouldn't have been so much interest in Congress in WIPP," Hancock said.[36] Energy officials took an understandably dimmer view of his efforts. "I don't know whether

Synar was environmental or it was just a case of Democrats versus Republicans," Duffy recalled. "Outside the committee hearings, he was a fine guy. But he had a message he was trying to get across."[37]

Just before the public hearings on the environmental study got under way in New Mexico in June, Synar's subcommittee held a hearing to review WIPP's status. Synar accused the department of "inviting failure" by planning to open in September. "Overall, the WIPP program seems to be guided by the philosophy of 'do it quickly' rather than 'do it right,'" he said. "Invariably, deadlines are missed, critical issues are ignored. Unanticipated problems are solved in a patchwork fashion which often gives rise to new questions and potential problems."[38] Synar produced a series of internal documents that he contended showed severe defects in the plant's management. Among those he cited was a memorandum by James P. Knight, director of Energy's Office of Safety Appraisals, who said that "significant additional effort in ongoing staffing, training, procedure development and documentation is necessary" before a recommendation should be made to bring waste for testing. The congressman also criticized the department's record keeping and noted the agency was months behind its deadline for completing detailed diagrams of 21 systems in the repository, including the electrical system, radiation control system, and fire protection system. And he noted that the responsibility for facility design appeared to rest with one man, Bechtel, Inc., engineer Howard Taylor. "It sounds like this whole $700 million project could go down the tubes if, God forbid, Howard Taylor ever disappeared," Synar noted.[39]

Such hearings were not the first time that Synar had waded so splashily into the WIPP debate. Nine months earlier, he made public internal reports at another hearing that showed WIPP's engineers were not yet certain the repository could be operated safely. "We're a month away from opening, and there are more Energy Department people lobbying for authority to emplace wastes in the repository than there are making sure the facility is safe," he said at the time.[40] Synar's method may have differed from the speakers at the WIPP hearings, but his purpose was the same: to attack expediency and add to the aura of scientific doubt.

Figure 22: Oklahoma representative Mike Synar (Congressional Quarterly *photo*).

"We Don't Need This"

Environmentalists and Synar were far from being the department's only worries. On the morning of June 6, FBI and EPA agents staged an unusual surprise raid of Rocky Flats, searching for violations of laws regulating the handling of waste that would be sent to WIPP. The allegations included illegal treatment and disposal of hazardous and mixed wastes, discharging pollutants without a permit or in excess of permit limits, false certification of federal environmental reporting requirements, and attempting to hide environmental contamination. The FBI said in an affidavit that infrared aerial photographs showed illegal waste incineration being done in a building that had been ordered shut down. The investigation eventually led Rockwell International Corporation, which managed Rocky Flats, to plead guilty to 10 charges of violating federal environmental laws. It was fined $18.5 million.[41]

Watkins and Duffy had been informed of the FBI raid, arguing unsuccessfully with the Justice Department not to use such a public show of force.
By coincidence, the raid came as WIPP's managers were attending the hearing
on the environmental study in Denver. The news caught them off guard.
"We don't need this," Marquez sighed.[42] Watkins, for his part, reacted
quickly, signing an agreement with the state of Colorado later that month
to spend $1.8 million a year on environmental monitoring and studies of
the health effects of plant operations. He also signed agreements to comply
with land disposal restrictions on solid and hazardous waste at Rocky
Flats. And he would subsequently cancel Rockwell International
Corporation's contract to run the plant and give it to EG&G, Inc.[43]

But the FBI raid was a severe blow to the department's credibility.
Not only did it give critics of WIPP additional fodder that the rest of the
federal government could not trust the department, it raised even higher
the bar of public confidence that the agency had to clear. Such perception
problems were not lost on Watkins. Three weeks after the raid, he announced
a new 10-point plan to shore up health and safety concerns at all Energy
Department sites. As part of that plan, he announced what others had
suspected: WIPP would not open in September. The admiral promised
not to set a new target date until he personally reviewed the entire project.
"I will not be driven by any previously set schedules or management
decisions which still do not answer emerging questions as to the soundness
of technical data or completeness of reviews," he said in a statement.
"WIPP is a classic example of the crying need to reestablish a well-aired
and documented baseline of understanding."[44]

New Mexico's congressional delegation welcomed the decision, noting
that technical questions continued to plague the project. "The fact that he
has firmly delayed opening WIPP until at least next year and has resisted
setting an opening date shows he's committed to answering those questions
in a thorough, deliberative manner," Senator Jeff Bingaman said. Environmentalists remained wary. "Hopefully, now he will agree with the
environmental community on just what 'safe' is," Kassen said.[45]

To underscore his commitment to the project, Watkins traveled to
Carlsbad the next day and met with Colorado governor Roy Romer and
Idaho governor Cecil Andrus. Accompanying them were an assortment

of Energy and Sandia engineers, who were struck by his pointed questions. Some of them believed that at last they had a boss who not only understood their technical work but was willing to fight for it politically. Romer and Andrus also were impressed. The two governors, who had triggered such a confrontation only nine months earlier, indicated they were willing to heed the secretary's approach. That evening, Watkins appeared at a dinner in Albuquerque with agency employees and civic officials intended to boost morale. He shouted, "I love WIPP!"[46]

Two months later, in late August, two boxcars of Rocky Flats transuranic waste reached their destination at Idaho National Engineering Laboratory. As Andrus had indicated during the Salt Lake City meeting in December, they would be the last shipments that his state would accept from Colorado. With WIPP not yet ready, a search began for alternatives. An internal task force looked at possible interim sites but met with predictable not-in-my-backyard opposition when details leaked out of some of the areas being studied. Among those who criticized the idea was New Mexico governor Garrey Carruthers, who told the agency that his state was already doing its part. "I didn't see any reason why we should help Andrus out and just clutter up our environment," Carruthers recalled.[47] Watkins himself seemed to recognize the dilemma. During his June visit to New Mexico, he acknowledged, "No one on Capitol Hill will talk to us about interim sites until we have an opening plan" for WIPP.[48]

The lack of political support for interim storage failed to deter others. Two companies that served as Energy contractors, Pacific Nuclear Systems of Washington and Roy F. Weston of Pennsylvania, in October proposed a temporary storage complex for Rocky Flats waste near Aguilar, Colorado, 30 miles north of the New Mexico border. The idea came from Miller Hudson, a former Colorado state legislator from Denver. Despite Hudson's attempts to sell the idea as an economic development project for a distressed area, it triggered vehement local opposition. It also sparked calls from Colorado lawmakers in Congress to push for an administrative land withdrawal that would allow immediate shipments to WIPP—something Watkins was not yet ready to try. Before long, the Aguilar proposal was dropped, although Hudson and some of its adherents subsequently began working with the Mescalero Apache Indian tribe of southern New Mexico on options for temporary storage of high-level, non-WIPP waste before Yucca Mountain opened. That proposal also ran into strong opposition and was killed.[49]

By the end of 1989, it was clear the chances were remote of another western waste storage political showdown like the one Andrus had touched off a year earlier. Weapons production had declined, removing the national security impetus the agency had raised at the Salt Lake City meeting. Instead the focus had shifted to Rocky Flats' problems. The Advisory Committee on Nuclear Facility Safety, an independent group charged with monitoring weapons production sites, issued a report in November giving an unusually harsh critique of Rocky Flats' operations. The next day, Watkins traveled to Colorado to announce that production at the plant would be halted indefinitely. "We haven't asked the right questions for too many years," the secretary acknowledged. "We did not bring our management into the modern world."[50]

As he apologized for past practices, Watkins began to implement some of the corrective changes he had promised Congress he would make. He established a new Office of Environmental Restoration and Waste Management that consolidated cleanup, compliance, and waste management activities within one office and named Duffy as its director. He also unveiled his five-year, $19.5 billion plan that prioritized cleanup activities. Critics said the plan did not go far enough in accomplishing their aims but at least offered a promising start. Finally, Watkins continued to impress on governors his belief that WIPP would herald the initiation of his new approach to how the government would clean up after itself. After the Rocky Flats raid and public hearings on the WIPP environmental study, his office took over control of the project from the Albuquerque Operations Office.[51]

When Colorado's Romer met with Watkins in December, he continued to have faith in the new secretary. "WIPP is still the solution we're all after," Romer said.[52] In the short term, the department's adherence to such sentiments helped to lessen its political troubles with Colorado and Idaho. But the method by which Watkins was to arrive at the so-called solution would trigger a larger and even more tangled series of interlocking intergovernmental and legal confrontations with environmentalists, Congress, and New Mexico state officials.

6

"No Discernible Scientific Basis"

1990–92: UNDERGROUND TESTING,
SCIENTIFIC DOUBT, AND CONGRESS

One morning in November 1991, Tom Udall tried to explain why he was suing the federal government to block the opening of the Waste Isolation Pilot Plant (WIPP). New Mexico's 43-year-old attorney general rummaged through stacks of documents until he unearthed a report that the federal government had filed to support its case in favor of sending nuclear waste to WIPP without waiting for Congress to act. The report was written by a member of a "blue-ribbon panel" that assessed bringing waste before the facility met the Environmental Protection Agency's (EPA's) long-term disposal regulations governing safety for 10,000 years. The question of whether the plant had to meet those standards remained the major obstacle confronting the Department of Energy (DOE) in its now 15-year quest to move materials into New Mexico. The report contained a sentence Udall had circled and highlighted in green ink: "I am concerned at what seems to be an obsession by DOE to get the first bins [of waste] emplaced in the underground." Udall pointed at the word "obsession" and smiled. "That explains a lot of why we're in this confrontation today," he said. "I think it's a pretty damning statement."[1]

Whether or not it was an obsession, the department's desire to use WIPP without authority from Capitol Hill set up what initially seemed to be a David-versus-Goliath struggle. On one side was Udall, a handful of environmental groups, and the state of Texas, who saw the department as overstepping its legal bounds in seeking to administratively rather than legislatively acquire land at the site from the Interior Department. They argued that the federal government could not prove the underground

Figure 23: New Mexico attorney general Tom Udall, with California representative
George Miller in the background (courtesy Tom Udall).

mine's walls would not collapse on workers. Opposing them was Energy
Secretary James Watkins, who wanted to prove his department was capable
of reversing the legacy of decades of weapons production problems that had
badly eroded its image. He had no trouble getting the Interior Department
to give him what he wanted and was anxious for the trucks to roll.

The legal victory would go to Udall, illustrating the power of the courts
to undo the actions of other branches of government. His lawsuit would
nullify Watkins's effort to bypass Congress, forcing the Energy secretary
to wait more than a year for lawmakers to agree on "land withdrawal"
legislation. In the process, the department's credibility further unraveled
in the eyes of its critics. The prestigious National Academy of Sciences,
which had backed underground tests with waste, became increasingly
skeptical and reversed itself. Congress imposed a set of conditions on the
tests that would prove impossible to meet before Watkins left office. In
all, the sequence of events demonstrated that no longer could any single
interest—such as the Energy Department—dictate WIPP's direction. Too
many branches of government, and outside groups with contrasting
agendas, were involved in the process for a cabinet secretary to declare it
was in "the national interest" to proceed.

The unfolding developments also showed that while political considerations drove the agenda at WIPP, a climate of scientific doubt remained

effective in a judicial setting. Udall's emergence as the official willing to take the Energy Department to court—a decade after Jeff Bingaman had first done so—solidified the belief among New Mexicans that they could employ legal means to lend a state voice to a federal project. The ambitious attorney general's challenge was an initial highlight in a career begun in the shadow of a famous family of environmentalist politicians. Udall's father, Stewart, served in the House and as Interior secretary under Presidents Kennedy and Johnson, where he wrote some of the nation's bedrock environmental laws, including the Wilderness Act of 1964. His uncle, Morris, was an even better known Democrat from Arizona who served as chairman of the House Interior Committee and ran for president in 1976.

After working as an assistant U.S. attorney in Albuquerque, Udall twice ran unsuccessfully for Congress. During his second campaign against Steve Schiff in 1988, he backed the need for the facility to meet the EPA's long-term standards, contrary to the Energy Department's belief that the plant could meet those standards after receiving waste for tests. Following the loss to Schiff, Udall worked with his father on behalf of two groups of radiation victims: Navajo Indians who had been exposed while mining uranium for use in nuclear weapons work and southern Utah residents living downwind of bomb fallout from the Nevada Test Site. The Utah residents who inhaled radioactive debris were shown to have unusually excessive levels of cancer, especially for a predominantly Mormon population that did not smoke or drink. Before health researchers documented those levels, the Energy Department dismissed the reports.[2]

"Part of the reason why I bring something to this issue that other people don't is my experience working with my father on the radiation cases and seeing the lies told to the downwinders and the uranium miners and how they were used as guinea pigs," Udall said. "I saw that the DOE and Atomic Energy Commission had a penchant for secrecy, saying, 'Because we're the ones who know the most about it, we know what's best.'"[3] In 1990, he entered the attorney general's race and won.

"Slim Just Left Town"

Half a continent away, Watkins scoffed at the notion that he was acting autocratically. He recognized the importance of building public confidence, forming a special task force to examine how to strengthen trust in nuclear waste programs. But the retired navy admiral continued to exert an unprecedented degree of control over WIPP. His approach to preparing for its opening was compared to a military exercise.[4]

Just as he had promised in August 1989, Watkins set up special teams to review every document and develop a database on the repository's design, engineering, construction, and operations. Department officials began issuing "draft decision plans"—documents containing so many criss-crossing arrows and symbols that they looked like the circuitry diagrams for a television set. Each delineated "technical/internal," "technical/external," and "institutional" requirements, with goals for each—including a projected opening date. Critics derided the emphasis on schedule setting. "Schedules are the stuff politics is made of," said Don Hancock of Albuquerque's Southwest Research and Information Center, the main field general in the environmentalists' unending war against the project. "It undermines public confidence when the only thing that matters is deadlines."[5] Energy officials responded that there was no other way to manage a sophisticated operation under so much political scrutiny. "You've got to have a basic schedule—if you don't, everything else is meaningless," said John Arthur, a manager at the department's Albuquerque office. "In this business, if you're going to go to Congress, you need to put together something realistic."[6]

But Congress, particularly the small but tenacious New Mexico delegation, was not about to be swayed by any schedule apart from its own. In April, Watkins sent the Senate a proposed land withdrawal bill that drew negative reviews from delegation members. The legislation contained few safety considerations that lawmakers sought; it placed no limit on the amount of waste that could be brought for testing. "It is clear that the administration bill is improper," Senator Jeff Bingaman said. Republican Pete Domenici outlined a series of demands before he would offer his support. His list included completion of a final report analyzing WIPP's safety, a plan for retrieving waste if the project was deemed unsuitable, and the purchase of a private potash lease on the site.[7]

Domenici's biggest demand, though, was money. He sought at least $200 million to build roads allowing WIPP trucks to bypass several cities throughout the state, including Carlsbad and Santa Fe—an amount Watkins considered extravagant. "Already we've got a real problem here, senator, in trying to find a source for these dollars in a very tough time," Watkins told Domenici at a hearing in May. An agitated Domenici pointed to a copy of the department's 1987 agreement with the attorney general to fund road improvements. "I think that you do not feel that you are bound by that," he thundered.[8] Such tense exchanges left proponents resigned to further delays. "The possibility of opening this year is slim to none," Representative Joe Skeen said. "And as one of the staff members said, slim just left town."[9]

By taking an adversarial stance with the New Mexico delegation over the road funds, Watkins lost an early chance to enlist the politicians who were most crucial to his goal of getting legislation passed. Perhaps more than anything else, compensation for highways united the state's disparate pro- and anti-WIPP factions. Politically, it provided politicians such as Bingaman and Representative Bill Richardson with a way to counter "not in my backyard" (NIMBY) sentiments against WIPP. At the same time, it gave staunch supporters a bargaining chip with which they could try to squeeze more money out of the federal government. "The noise about transportation was more of a leverage to acquire funds to improve roads than it was a deep-hearted sense of impending danger for hauling these things down the road," Governor Garrey Carruthers acknowledged after he left office in 1991.[10]

Senate Energy Committee chairman J. Bennett Johnston of Louisiana shared Watkins's desire for quick action, but the days when one politician could steer a WIPP bill into law had passed. New Mexico's delegation was now in a strong position: Domenici and Bingaman were respected members of the Energy Committee, and Richardson was close to the House Democratic leadership. All three were buoyed by polls showing a hardening of residents' attitudes toward the site. A University of New Mexico Institute for Public Policy report showed the number of respondents who believed the site was unsafe increased between August and November 1990. Overall, 59 percent of the respondents said WIPP was currently unsafe but that it could be made safe with either major changes (29 percent) or minor ones (30 percent). Another 27 percent said the site was unsafe and should never open. Just 10 percent believed WIPP was safe to open right away.[11]

Some department officials chalked up the delegation's stance to its desire to score political points. That WIPP did not open in 1990 "was basically attributed to Pete Domenici's running for reelection," waste management director Leo Duffy groused years later. But Domenici (who subsequently won a fourth term with ease) had a more practical reason for not acceding to Duffy's wishes. He had learned that the longer he held out on land withdrawal, the more incentive it gave the department to finish other technical tasks. That, in turn, helped build public confidence that the agency was not simply trying to get Congress to cut corners on its behalf. "It's leverage," the senator explained.[12]

As policy makers squabbled over highway funding, New Mexicans debated whether the bypasses were enough to eliminate fears about transportation. In February 1990, a jury in Santa Fe awarded a couple living along the city's planned bypass $337,815 because the value of their property was reduced by plans to use the road for shipments. The couple, John and Lemonia Komis, went to court after being unable to reach agreement with the state Highway and Transportation Department over compensation for their land. The jury was swayed by a public opinion survey presented by the Komises' attorney that found 59 percent of the respondents would not buy residential property along the waste transportation route. To critics, the decision showed that the judicial branch of government remained a place to turn.[13]

With the Energy Department unable to make any progress with Congress on land withdrawal legislation, it addressed other hurdles. One was the issue of hazardous wastes at WIPP under the 1984 amendment to the Resource Conservation and Recovery Act (RCRA). Much of the waste destined for the site was both radioactive and hazardous, containing items contaminated with materials such as lead and cadmium. The RCRA law classified certain materials, such as solvents and forms of industrial sludge, as so dangerous they could not be buried unless they were treated to reduce their toxicity. But it allowed for an exemption if it could be shown that burying the waste would not leak into underground water supplies for as long as they remained hazardous.

The Energy Department filed such a "no-migration" exemption for WIPP with the EPA in March 1989. The EPA granted it in November 1990, imposing several conditions. Among them was limiting the amount of hazardous waste to no more than 8,500 drums, or 1 percent of the plant's total capacity. In addition, officials called for testing the gases of each drum of waste to ensure they were not flammable. To Duffy, the EPA's action provided enough of a reason to open the plant. He also saw it as proof that the department was willing to submit itself to outside regulators. But environmentalists said the decision marked another example of the Energy Department's haste to bury waste, this time with the EPA's complicity. They noted that environmental regulators spent more money to process the petition in 1989 than they spent the same year on rewriting the much broader reaching radioactive waste disposal standards. "EPA is driven by DOE's plans for WIPP," Hancock said.[14]

After receiving the no-migration exemption, Watkins continued to press Congress to act on WIPP, describing it as his highest legislative priority. But Bingaman and Domenici would not budge. "Frankly, we can see no reason for DOE to push for land withdrawal legislation this year," the senators said in a letter, "except to fulfill an arbitrary timetable."[15]

"The End of Our Negotiating Rope"

In his dealings with New Mexico, Watkins repeatedly threatened to carry out an administrative withdrawal—a formal interagency legal agreement between the Interior and Energy departments—if Congress refused to act according to his schedule. As both sides were aware, such a move would deny the state any of the road monies or health and safety provisions included in legislation. Although Watkins promised to address the state's needs, New Mexico lawmakers countered there was no way to legally bind him or his successors.

An administrative transfer required the approval of the Interior secretary, who happened to be former New Mexico House member Manuel Lujan. Although Lujan had planned to retire in 1988, President-elect George Bush implored him at the last minute to become the overseer of one-third of the nation's land. Lujan's selection drew protests from environmentalists who suspected the affable, low-key Republican was picked because he was Hispanic. Lujan did little to discourage criticism in his early months in office, committing a series of embarrassing gaffes. He told Alaskans that the Exxon Valdez oil spill could boost tourism in their state and had to be corrected by reporters when he misinterpreted the controversial Mining Act of 1872.[16] As a congressman, Lujan backed former Democratic governor Toney Anaya's attempts to give the state greater control over WIPP but also supported the test plan. "It's low-level waste," Lujan said, committing an error often made by those misidentifying the longer-lived transuranic materials. "It can't blow up. It's not a Chernobyl-type situation."[17] When he became Interior secretary, some New Mexicans chastised him for not remaining loyal to his home state on WIPP. "I do feel some sort of loyalty," Lujan responded. "Having said that, I also feel a responsibility to this country."[18]

Lujan granted an administrative withdrawal in January 1991, following the Bureau of Land Management's recommendation to modify the 1983 public land order that had allowed construction on WIPP. The modification included an expansion of the withdrawal to conduct tests with "retrievable" radioactive waste for the planned six-year test phase. Duffy was satisfied legislative action was not needed and that it was, in essence, too late for the state to object. As he saw it, the land near Carlsbad already had been dedicated to WIPP—if New Mexico officials disagreed, they should have done so in the late 1970s when drilling teams struck underground brine pockets.[19]

Lujan remained hopeful his former colleagues in Congress could act. The lawmakers, in turn, did respond—though hardly in the way Watkins or Lujan wanted. At Richardson's urging, the House Interior Committee passed an emergency resolution in March that purported to nullify the administrative withdrawal. The action was largely symbolic but made clear Congress's intention it would have the final say. In a slap at one of his harshest critics, Richardson sent a copy of the measure to Republican activist John Dendahl, who had continued his newspaper-letter-writing campaign on the project's behalf. "This bill's for you," the lawmaker scrawled on it.[20] Despite Dendahl's protestations that Richardson was simply out to stop storage, the congressman insisted that was not the case. "While never a big fan of WIPP, I have always supported the project so long as we can be 100 percent certain that it can operate safely," Richardson wrote in a letter to the *Albuquerque Journal.*[21] Dendahl immediately challenged the statement. "The congressman ... knows that 100 percent safety cannot be achieved in any endeavor, but he has slyly hidden behind that qualification for many years," he responded.[22]

Such bickering illustrated the continued lack of consensus over how to assess "safety." Would WIPP be safe, as supporters contended, because it had been carefully built and managed, something they said putting waste underground would prove? Or, as environmentalists responded, was rigid external oversight and further bureaucratic approval required before a single shipment was allowed? The three House committees—Interior, Energy and Commerce, and Armed Services—struggled to find a middle ground.

The legislative action finally kicked off before the Interior panel, whose Energy and Environment Subcommittee approved a bill in June. The full committee followed suit two weeks later. The measure, sponsored by Pennsylvania Democrat Peter Kostmayer, was hardly to the Energy Department's liking: it required that the project be subject to EPA oversight during the test phase. The bill also placed a 10-year limit on the land transfer, a restriction that Interior Committee chairman George Miller, a California Democrat, said was needed to allay unease over the department's spotty track record at running its nuclear facilities. But in a snub to Richardson, it allowed underground tests with one-half of 1 percent of the facility's capacity, or about 4,250 drums. The reduction nevertheless marked a far cry from the 125,000 drums—15 percent—that the department had insisted it needed a few years earlier. One reporter dubbed the cutback "the Incredible Shrinking WIPP Test Plan."[23]

Richardson was forced to withdraw an amendment to have WIPP meet all EPA standards after it became clear he lacked enough votes; several Democrats said they were reluctant to give New Mexico so much control. Similarly, although they did not attempt to take out the $397 million authorized for highway bypasses and emergency preparedness plans, lawmakers worried that the money would set a harmful precedent. "I think we're starting down a course in which many congressional districts in this country can claim a right" to compensation, said Philip Sharp, an Indiana Democrat.[24]

As the months passed, Congress remained deadlocked. In the Senate,

Domenici and Bingaman tried to ignite some momentum, introducing a compromise that allowed 4,500 drums underground before the EPA's long-term disposal standards were met. But their bill also required the EPA, not the Energy Department, to certify compliance with the standards. It granted sweeping oversight powers to the state and Environmental Evaluation Group (EEG), who were given the task of reviewing and commenting on a plan for the tests along with the National Academy of Sciences. "This bill has in it more safeguards for New Mexico than have ever been tied to legislation before," Bingaman said.[25] The critics remained unhappy; Udall noted it did not allow for a judicial review by the courts.[26]

The debate on Capitol Hill no longer seemed to faze Watkins. In July, he told a reporter the plant would open before the end of the year—"a major achievement." His comments reflected his supreme confidence about surmounting a court challenge. "Sure, we will be sued," he said. "But we'll be sued not on the basis of environmental concerns. We've already demonstrated we know how to deal with all safety and environmental aspects of WIPP. The suits will be emotionally anti-nuclear—'We don't like bombs and we don't like anything connected with bombs, even safe waste disposal.' OK, I understand there is a body of American thinking that says the nuclear deterrent is bad. I don't think it is majority thinking. But only pure emotionalism from this faction stands between us and accepting wastes at WIPP."[27]

Two months later, the secretary demonstrated he was serious. The department certified on October 3 it was in compliance with all environmental requirements at the facility, leading Watkins to announce that waste would begin arriving as early as the following week. He cited a breakdown in negotiations with Domenici and Bingaman over how much could be brought for tests and pointed out that the site was costing taxpayers $13 million a month. "We've reached the end of our negotiating rope," he told reporters.[28]

Many New Mexicans felt betrayed. "In going ahead on an administrative basis ... the secretary needlessly causes conflict with our state and with the congressional delegation," Bingaman said. Angry activists discussed erecting barricades to block waste trucks.[29] Meanwhile, Nevada officials who were battling the department over Yucca Mountain for high-level

Figure 24: Room 1 of Panel 1, the underground area at WIPP originally planned to hold waste during the test phase (Department of Energy photo).

waste disposal said Watkins's action typified the agency's disdain for states. "It is another example of why Nevada should question any promises the DOE makes," said Democratic senator Harry Reid. "It shows that the DOE, when it deals with anything nuclear, is not concerned with state's rights."[30]

Building a Case Against Testing

Well before Watkins's announcement, Udall had been building a case against WIPP's test phase that did not rely on the "pure emotionalism" the secretary predicted would propel a lawsuit. He assigned several lawyers to the matter, including Lindsay Lovejoy, an assistant attorney general in his office's environmental enforcement division. Lovejoy came to New Mexico in 1977 to work on uranium contract litigation, spending three years at Bingaman's old law firm, then another decade on utility and natural resource issues. Lovejoy's skill in absorbing technical data impressed Udall, as did his doggedness in tracking project developments.

Lovejoy believed the department had created a problem in trying to bury radioactive materials in an area so abundant in natural resources.[31] He met with others to devise a legal strategy. They settled on the notion that WIPP was geologically unstable and waste could not be retrieved safely if brought underground. The salt walls of the rooms that had been mined in the early 1980s were shown to be closing in at a rate three times faster than anticipated. As those rooms reached the end of their expected life span, they began to collapse. Sections of ceiling weighing hundreds of tons cascaded down inside the rooms. In June 1990, workers were stunned to find a pile of rubble lying in one room where they had taken measurements a few days earlier. "We could have been in there when it happened, and thinking about it even now scares the hell out of me," Tia Mills, one of the workers, told The *New York Times* two years later. "Every time we went into that room, we discussed it among ourselves that we shouldn't be there, because when that stuff comes down, it doesn't give you any warning."[32]

At WIPP, the significance of roof falls was downplayed, with officials saying such collapses were anticipated. Most of them, they noted, occurred in experimental rooms that had been heated to study how salt eroded. Internally, though, the June incident caused a stir. An investigation showed rock gauges were picking up the faster rates of salt movement in the ceiling of the room, but no one had read the computer printouts for at least a month before the collapse. As a result, Sandia National Laboratories began reviewing the data daily.[33] The officials also developed a million-dollar engineering system. The system used rock bolts—13-foot metal rods anchoring sections of the ceilings to prevent roof falls. The bolts were electrically monitored and grouted with a special epoxy to increase stability. As an added precaution, a mat of steel mesh was placed beneath the bolts in the first room where the waste was to be sent. "We will be doing just about all that humanly can be done for that room," said Tod Burrington, an official with Westinghouse Electric Corporation, the department's WIPP contractor.[34]

Yet there remained no consensus such supports would work. Jack Parker, a Michigan geologist and mining engineer who had served on a panel to evaluate the collapses, provided the Attorney General's Office with an affidavit that said the roof of the room where waste was to be brought was unstable. He predicted it would fall before the testing ended and said the idea of putting waste in the room was "imprudent." Lovejoy brought in

Figure 25: A WIPP employee installs rock bolts in WIPP's ceiling underground (Department of Energy photo).

other outsiders to raise more doubts. The health and safety coordinator for the National Union of Hospital and Health Care Employees concluded that a review of New Mexico hospitals along the shipping routes found most of them unprepared to deal with any accidents involving a radiation release.[35]

The federal government countered with its own experts. In one affidavit filed to support the Energy Department's case, the chairman of the Advisory Committee on Nuclear Facility Safety, John Ahearne, said the first set of tests "can be conducted safely and there are no significant risks." The head of a radiation emergency response center in Tennessee, Robert Ricks, also refuted claims about hospitals' preparedness. The fact that no consensus existed made some New Mexico officials uneasy about being in court. "The bottom line is that it's difficult for a state to defeat the feds, especially in an area of law that the federal government says is part of a defense necessity," Steve Schiff said.[36]

But Udall and others initiating the litigation believed the law was on their side. They argued the Bureau of Land Management was given the responsibility of regulating a subject outside its jurisdiction and about which it knew nothing. Their lawsuit alleged several Interior Department violations of the Federal Land Policy and Management Act of 1976 (FLPMA). They included Interior's failure to determine that the proposed use of WIPP for underground tests would be completed within the time period of the withdrawal and its failure to show that the purpose for which the withdrawal was originally made required an extension for testing with waste. The suit also alleged violations of both FLPMA and the National Environmental Policy Act (NEPA) based on Interior's policy reversals to allow radioactive waste testing under an administrative withdrawal.[37]

Three weeks after Udall filed for a temporary restraining order and preliminary injunction in U.S. District Court, several other parties joined the case. They included four environmental groups: Southwest Research, Concerned Citizens for Nuclear Safety, the Environmental Defense Fund, and the Natural Resources Defense Council. Three Democrats on the House Interior Committee—Richardson, Kostmayer, and Utah representative Wayne Owens—also signed on as plaintiffs. Finally, the Texas Attorney General's Office came aboard, reiterating fears about contamination reaching the Pecos River. The resulting combination of state, federal, and citizen interests reassured Udall that the opposition was broad based enough to show the suit stemmed from more than any NIMBY fears.

The judge assigned to hear their arguments was John Garrett Penn, a 59-year-old Massachusetts native. Penn served in the Justice Department during President John F. Kennedy's administration and won an appointment from President Richard Nixon to a newly created Superior Court for the District of Columbia. He spent a decade hearing criminal cases until President

Jimmy Carter appointed him to district court in 1979, making him one of the court's first black judges. As a teenager, Penn loved science, and he majored in chemistry during his first two years in college. His familiarity with technical issues, and record in past cases, left Udall optimistic.[38]

Penn held a hearing on the request for a preliminary injunction on November 15. In his opening remarks, Udall charged that "a decision was made to not follow the law" in Watkins's desire to proceed with an administrative withdrawal. The attorney general outlined his concerns over the stability of WIPP's test rooms, noting the risks of roof falls and the department's unwillingness to consider other locations. Justice Department lawyers representing the departments of Energy and Interior refuted Udall's legal points. One of the attorneys, Caroline Zander, pointed to the engineering enhancements in place and the ability to predict roof falls in advance. "The bottom line here is that nothing is irreversible," Zander said, describing the opponents' case as consisting of "theatrical scenarios" based on speculation. Penn, however, appeared to pick up on Udall's reasoning. He questioned the claim that the Energy Department needed to use the plant right away. "Why don't you just wait for congressional action?" he asked. Zander responded that Congress had been given four years to act. "It's impossible for us to use a crystal ball," she said.[39]

The judge was expected to reach a decision in a month but ended up taking 10 days. He issued a memorandum opinion and order granting a preliminary injunction, blocking any shipments. Penn found the Interior Department made no determination that the purpose for its 1983 land withdrawal would require an extension to include underground tests with waste. He ruled that the extension altered the original withdrawal in such a way that it was outside the bounds set forth in FLPMA, under which an extension is necessitated by its original purpose. Penn also said the government presented "no convincing evidence" the waste could be safely retrieved and rejected Watkins's claim that WIPP should open because it cost taxpayers $13 million a month to maintain. That money, he noted, would be spent regardless of whether waste was brought. Udall had gotten what he wanted. "This is clearly a major victory for New Mexico," he exulted.[40]

Other state officials and politicians expressed hope that Penn's ruling would give Congress time to act. But as news of it spread, lawmakers in the House were hours away from adjourning for the year and still stymied over how to reach a deal. Prodded by the lawsuit, the chairmen of the Interior, Energy and Commerce, and Armed Services committees agreed to temporarily put aside their differences and send the Interior-approved bill sponsored by Kostmayer directly to the House floor. But Sharp, the cautious chairman of Energy and Commerce's Energy and Power Subcommittee, balked at allowing a vote, fearing that the House—restless as it rushed toward the adjournment—might defeat the measure. Sharp wanted more time to line up votes.[41] The bottom line: WIPP would have to wait.

National Academy Panelists Blast the Tests

The path to passing a bill in 1992 was strewn with obstacles. It was the final year of the Bush administration and an election year, a time when Congress does relatively little. Both sides had staked out their positions to the point that achieving an agreement proved difficult, particularly with outside interests—the Energy Department on one side and environmentalists on the other—applying pressure at each end. The department had made such a large investment in the idea that it needed to bring in waste for tests, it was prepared to fight any challenge. By the same token, environmental groups were hardly about to let radioactive materials be sent to a facility that had been legally judged to be unprepared.

The Senate had done its part to move things along. It passed the Domenici-Bingaman bill in November, a week before Penn's ruling. Environmentalists disliked the measure, though the department could live with it. It allowed burial of 4,250 drums—one-half of 1 percent of overall capacity—during the test phase, with further shipments pending the EPA's approval. Bingaman tried to get the Energy Committee to adopt an amendment capping storage at 1 percent, but conservative Democrats Johnston and Richard Shelby of Alabama joined with the panel's nine Republicans to defeat it. Johnston argued that Bingaman "risked losing the whole ballgame" if the amendment was adopted, because the Bush administration would abandon its support of the bill.[42]

House members received another incentive in early February, two months after Penn's initial ruling, when the judge made permanent the preliminary injunction. The same day, the judge also ruled on separate lawsuits that had been brought by environmentalists over hazardous waste storage. He blocked Watkins from putting any waste at WIPP until the department completed the process of obtaining a hazardous waste permit from New Mexico. The legal situation led Andrus to warn of looming "negative impacts" to Idaho and other states.[43] Watkins portrayed the situation as even more urgent. He cautioned that layoffs could occur at the WIPP site if the House took up the Interior Committee version, saying its provisions could delay opening up to a year. At the time, the department was keeping WIPP fully staffed in anticipation of the arrival of waste, paying more than $400,000 in overtime to workers between September 1991 and February 1992.[44]

Penn's decision to issue a permanent injunction complicated the Justice Department's efforts to appeal the initial ruling. The department did file with the D.C. Circuit Court of Appeals, but that court quickly affirmed Penn's decision. It rejected the argument that the Interior Department's decision for a "modification" of its original land withdrawal would only justify a change in the purpose of an original withdrawal—not an extension. Consequently, the appeals court said, an extension could not be granted to accomplish the purpose of a modification.[45] Duffy was unmoved, blaming the Justice Department's lawyers for not emphasizing the need for testing to meet the EPA's RCRA requirements. "We had three weeks of discussions, and they couldn't come up to speed on what [RCRA] had to do with the WIPP operation and the no-migration variance," he said. The appeals court did uphold the department on RCRA: it said the plant had "interim status" under the law and could receive waste without a state permit. Under interim status, hazardous waste facilities could operate while applications for state permits were pending.[46]

Watkins remained as anxious as Duffy. "We ought to open that plant and start the research work," the secretary said in January. "That research is important. If it proves out, it's much safer to store the transuranics underground in that salt mine than where they are today. . . . We have had five to seven external review groups totally independent of me say we're ready."[47]

However, the most distinguished review group was now having its doubts. The National Academy of Sciences' Panel on the Waste Isolation Pilot Plant, made up of a dozen geologists, engineers, and other prominent scientists, had previously gone along with the department's assertion that testing at WIPP was needed. Department officials badly wanted the panel to remain in their corner; it was the academy, after all, that had proposed in 1957 that salt be studied as the best medium for burying radioactive materials. At a December 1991 meeting, they beseeched the panelists to back them up. "The future of WIPP depends on your support," Mark Frei, director of the department's waste management projects division, told them. "Without it, we could be in court for years, or worse, have to terminate the project."[48]

The department was seeking to conduct the tests in two phases. The first, called the "bin tests," involved placing plutonium-contaminated items inside rectangular metal containers, or bins, to measure how much pressurized gas was released. A second and more extensive set of experiments known as "alcove tests" called for storing waste drums inside underground rooms to study their behavior. The Energy Department first discussed measuring gas generation in October 1987, proposing to fill up four rooms with 24,000 drums. Because of deficiencies, that plan was abandoned. Two years later, in April 1990, the department issued a new outline of the bin and alcove tests.[49] Energy Department officials argued the tests were needed because they would help understand how much gas the buried wastes would generate. That, in turn, would help them meet the EPA's long-term disposal standards. Lawmakers such as Domenici endorsed the idea because they felt putting small amounts of waste underground would help raise public and political confidence.

But New Mexico EEG members, especially deputy director Lokesh Chaturvedi, stepped up arguments that the experiments would only confirm what already was known about gas generation, pointing out the tests did not have to be done at WIPP. Department officials claimed it would take too much time and money to develop facilities at one of the sites where the wastes were stored. But as Chaturvedi noted, if an effort to prepare laboratories at Idaho National Engineering Laboratory or Rocky Flats had begun in 1989, the department could have been getting data by 1991.[50]

At the academy's December meeting, panel members raised similar concerns. "We're not being asked to approve a set of experiments; we're being asked to approve a process of experiments for an open-ended period of time," complained panel member John Bredehoeft, a California geological engineer. Panelist B. John Garrick, another California engineer, told Sandia officials that panel members "are still struggling with what knowledge or incremental value these bin tests will provide." After the meeting, though, panelists agreed to continue supporting the tests. The panel's chairman, University of Minnesota engineering professor Charles Fairhurst, explained that bringing a few bins would help the agency check out its transportation and waste-handling systems and would not pose a significant safety threat.[51]

In the following months, however, Fairhurst and the others decided they could no longer defend that position. In a startling development, they sent Duffy a five-page report in June that bluntly criticized the need for testing with waste at WIPP, echoing much of what critics had been saying for years. "The panel has not been convinced by the scientific rationale, as presented, for the underground gas generation tests," the report said. It noted that a series of bin tests using "dry" materials such as glassware and tools was a "serious misallocation of resources" and that those tests were unlikely to produce much useful data. Although it did not reject the idea of doing any underground testing, the report concluded: "The plan to conduct a large number of expensive bin tests and to terminate the experiments after five years has no discernible scientific basis."[52]

The academy's findings represented the ultimate example of how scientific conflicts could fuel political controversy on WIPP. For the first time, an influential independent group that had backed the department's plans had given every appearance of reversing course. The result prompted intense news coverage. Watkins and Duffy were furious, fearing it could kill any chance of congressional action. The two had joined several governors—including Andrus, King, Colorado's Roy Romer, and Nevada's Bob Miller—in lobbying to get the House to take up a land withdrawal measure. They had even flown the governors to Carlsbad, hoping to present a united front. "I've never seen a collective activity of man that has been so cautious and almost so overengineered," Romer marveled after the tour.[53] But Watkins startled King during the visit by publicly suggesting that the New Mexico governor had the power to open the plant to allow testing under RCRA. King later issued a statement that it was Congress's call, not his.

Watkins and Duffy launched a damage-control effort in response to the academy's report. Duffy urged Academy of Sciences president Frank Press to "clarify" the panel's statement on the tests. The next day, Press wrote to the chairmen of the three House committees with jurisdiction over WIPP to contend that some newspaper accounts of the academy report had "misinterpreted the panel's findings." Press concluded, "I wish to assure you of the panel's continued support for an underground testing program at WIPP."[54] Panel members were livid. WIPP supporters asked panelist Christopher Whipple, a California environmental consultant, to sign a letter to *The Washington Post* "that made everything all right" and advocated underground testing. He refused. "They had one drafted for me, and I said I couldn't do it," he said.[55]

In justifying his actions, Duffy argued that the academy panelists—like the Justice lawyers—failed to address the need to bring waste under RCRA. "They did not understand the RCRA portion of the operation," he said later. "They were not even remotely associated with the no-migration variance."[56] Watkins, meanwhile, saw the issue as a matter of using an expensive facility whose fragile walls were crumbling. "We could have done the tests anywhere, yeah," he recalled. "But the salt was coming in at WIPP."[57] To critics, such arguments were unconvincing. Richardson charged that Press's attempt to pacify Congress represented "a turnaround at the pressuring of the Department of Energy."[58]

At Long Last, a Bill Becomes Law

Four weeks after the furor over the academy's report, the House finally took up land withdrawal legislation. The measure represented a compromise among the three committees but was generally closest to the Energy and Commerce–approved bill. That bill was a middle ground between the Interior measure, which environmentalists preferred, and the Armed Services bill, which Watkins favored. The three chairmen were satisfied with the compromise and pledged to oppose any amendments to it.

Richardson was ready for a fight. "As we open the first DOE facility in

Figure 26: *Idaho governor Cecil Andrus (second from right) at WIPP with, from left, Energy Secretary James Watkins, Nevada governor Bob Miller, and Leo Duffy (Department of Energy photo).*

30 years, do we trust DOE to manage this facility with all the safety, health and environmental oversight that is required?" he asked during debate. "The answer is a resounding 'no.'"[59] But the congressman lacked the clout of the committee chairmen. All were prepared to accede to the department's insistence that it bring what was regarded as a relatively trivial amount of waste before it opened for long-term disposal. By and large, they were satisfied the bill contained enough safeguards to ensure the department would not operate in a completely unregulated fashion. And perhaps most important, they saw in the plant an opportunity to put Congress's stamp on the widely publicized problems that plagued the nuclear weapons complex. "What the critics fail to acknowledge is that the waste that is designated to go to WIPP is currently stored, unprotected, in warehouses all over the nation," George Miller said. "As a society, we must begin to rectify this situation."[60]

Joining the chairmen were lawmakers from the 10 states where the waste destined for WIPP was stored. Democrat John Spratt of South Carolina, an influential Armed Services member, echoed Watkins's argument that it was time to see if the plant would work. "Now that we have spent $1.2 billion to build this facility, it seems only logical to use WIPP itself to determine whether or not WIPP can do what it was designed to do," he said. Even Oklahoma Democrat Mike Synar, one of Energy's severest critics, spoke in favor of the bill, pointing to the reduction in the amount to be shipped. Initially the tests "were a pretext for opening WIPP and resolving the severe transuranic waste storage problems DOE was experiencing as a result of the state of Idaho's refusal to accept more waste," Synar said. "To its credit, DOE has largely abandoned these transparent efforts to open WIPP without meeting environmental standards."[61]

As a result, Richardson's amendment to require that both the EPA short-term and long-term standards be met before waste was brought was defeated, 253–148. Although many liberal Democrats and some moderate Republicans supported it, they were outnumbered. The final bill passed 382–10, with Richardson among the six Republicans and four Democrats in opposition. The action set the stage for a conference with the Senate to work out the differences between the bills. Although such conferences took place behind closed doors and the public paid little attention, it was where the real business of legislating was done.

Achieving a compromise took nearly three months. During that time, environmentalists continued to pressure Domenici and Bingaman to support the safety provisions in the House version, taking out full-page newspaper advertisements asking them to "stand up for WIPP." The environmentalists tried to ensure the tests would be done only if they could be shown to be scientifically valid through the EPA process. The House bill said the tests had to be "necessary to demonstrate compliance" with the standards, while the Senate bill said they should "provide relevant and useful data in a timely manner." The department preferred the less restrictive language of the latter.[62]

The final agreement ended up requiring the tests to be "directly relevant to a certification of compliance." It gave House negotiators many of the other safety provisions they had sought but also sped up the timetable for federal action, which pleased Senate conferees and agency officials. Most important, the EPA remained the independent regulator of WIPP. The agreement called on the agency to issue new standards within six months of enactment and stated that the agency's approval of the testing plan had to come within 10 months. It also required the Energy Department to issue plans for retrieving the waste and decommissioning the facility. It barred high-level waste, even for experiments. As for compensating New Mexico, the agreement included an annual $20 million payment for 15 years for highway bypasses and other costs, which was almost half the $600 million in the Senate bill but significantly more than the $40 million in the original House measure. "This is a good resolution of a very difficult series of questions, and a very, very large number of conflicting interests and concerns," said House Energy and Commerce chairman John Dingell of Michigan.[63]

The House approved the conference report without taking a roll-call vote on October 6, with the Senate following two days later. Bush signed the measure into law on October 30, days before he would lose his bid for reelection to Arkansas governor Bill Clinton. After more than five years, the largest legislative hurdle facing the plant had been cleared; Carlsbad residents were relieved. "When WIPP was built, we were told there would be no problem withdrawing the land—it was to be just a formality, since the project was built on federal land," said former mayor Walter Gerrells. "The overwhelming vote in Congress shows that, as a whole, they are supportive of the project."[64]

The enactment also pleased Watkins, though he would be unable to give the signal to start the trucks rolling. The earliest that waste could be put underground was now August 1993. Years later, he lamented the degree to which dueling science had thwarted him in his attempts at fulfilling an orderly process. "I tried as hard as anybody could try to get that thing open," he said. "Yes, you could speculate that 10,000 years from now, someone could drill down and get shot in the eye [with waste]. Nobody has been able to meet engineering standards for 10,000 years. So you end up in those situations where all you hear is, 'Yeah, but what if?' We did all the things that were required of us, but it didn't make any difference."[65]

Predictably, Udall saw the situation differently. To him, Watkins had succeeded in cutting scientific corners to achieve a political end on a questionable approach to using WIPP. "The whole issue of the tests seemed to me that what they were trying to do was get their foot in the door," he said. "It really wasn't based on science; they could have done it much more cheaply in other locations. It kind of troubled me they were pushing this test phase so much, yet it didn't seem to have much basis."[66] As it would happen, a new Energy secretary would take the project in a radically new direction—but she would remain unable to end all conflicts with environmentalists and New Mexico officials.

7

"A Major Break"

1993–96: CANCELING TESTING, PREPARING FOR DISPOSAL

Hazel R. O'Leary disliked many things at the Department of Energy, including the jargon-laced way its employees talked. Shortly after becoming President Bill Clinton's secretary of Energy in January 1993, she began slapping 25-cent fines against anyone she caught using acronyms such as FFTF or NPR instead of Fast Flux Test Facility and New Production Reactor. The proceeds—more than $80 within a few months—went to the department's child development center. To O'Leary, the fines were a small signal that the traditional approach of operating no longer applied. "It was one of the many elements I thought we could change about the way we conducted our business, which was to pull people in to understand what we're all about," she recalled.[1]

When it came to the Waste Isolation Pilot Plant (WIPP), O'Leary championed substantive change. An initial advocate of underground testing, she soon reversed course and scrapped the idea to concentrate on winning Environmental Protection Agency (EPA) certification of the site. At the same time, O'Leary made other moves that flew in the face of what her predecessor James Watkins had done. She ceded control of the project's operations to a new office in Carlsbad. She placed a new emphasis on reaching out to affected interests, or "stakeholders." She even hired some of the environmentalists who had fought the department on WIPP. Although O'Leary's political fortunes eventually sank, WIPP was put in position under her tenure to receive the EPA's subsequent stamp of approval. Nevertheless, the efforts of O'Leary and others did not succeed in clearing all of the other obstacles necessary for waste shipments to start. Nor did they remove all criticism over the department's waste management policies. As Watkins had learned, the pressure from outside groups and New Mexico's attorney general made it impossible for the department to proceed without encountering major criticism and delay-inducing legal challenges. Even if those critics could not stop WIPP outright, they could slow it down.

Nowhere was that clearer than with the EPA's review of WIPP. Far from resolving all disputes over safety, the certification effort again showed the collision of science and politics. Such a process was an inevitable outcome of an independent regulator's subjective effort to guarantee the plant's safety for centuries, well past a period in which any outcome could be certifiably determined. Congressional Republicans, mistrustful of the EPA, charged that the environmental agency took too much time and was overly stringent, and they passed legislation aimed at speeding up the process. Meanwhile environmentalists and New Mexico attorney general Tom Udall tenaciously took the opposite tack. They argued that the EPA cut corners and made major changes in its criteria for determining the plant's safety, prevented public participation, and withheld critical technical reports. They also charged that the EPA was far too timid in its dealings with Energy, with one of them likening its oversight to "Bambi riding herd over Godzilla."[2] They turned to the place that had now come to resolve most WIPP-related disputes—the courts.

O'Leary, 56, had not been Clinton's first choice for Energy secretary. Despite having held energy-related jobs in business and government, she lacked experience in nuclear weapons cleanup. She had spent three years as an executive at Minnesota's Northern States Power Company and had been promoted to president of its subsidiary gas company. As a federal employee, she ran the Economic Regulatory Administration for President Jimmy Carter and handled consumer affairs for the Federal Energy Administration under President Gerald Ford. Because O'Leary was an African American woman, some Clinton supporters saw her selection as proof of his commitment to a cabinet that "looks like America." Announcing her appointment, the president-elect stressed her sensitivity "to one of the biggest problems the Department of Energy has, which is that very often on its most controversial decisions, it has very little credibility."[3]

Figure 27: *Sandia's Wendell Weart talks to, from left, Senator Jeff Bingaman, Energy Secretary Hazel O'Leary, Representative Joe Skeen, and WIPP Manager George Dials (Department of Energy photo).*

Echoing such sentiments, the Task Force on Radioactive Waste Management issued a subsequent report painting a dismal picture of the department's public trust and confidence problems. Despite Watkins's efforts, the report found "a widespread lack of trust" in nuclear waste management that it said was neither irrational nor a manifestation of not-in-my-backyard (NIMBY) syndrome. The task force compared WIPP with both the civilian radioactive waste management program and the cleanup of contaminated weapons production sites. It found that the civilian program faced greater obstacles to recovering and sustaining trustworthiness than the cleanup effort, while WIPP "generally occupies a position between the other two." It foresaw a tough road ahead for the department to win over critics: "This will continue for a long time, will require sustained commitments from successive secretaries of Energy to overcome and will demand that DOE act in ways that are unnecessary for organizations that have sustained trust and confidence."[4]

O'Leary set to work on such problems in the manner of a corporate

CEO. She began emphasizing diversity and "a new spirit of inclusiveness, communication, and openness" to give workers "a sense of mission ... and pride." She brought in management guru Stephen Covey to address her senior staff; employees grew accustomed to buzzwords such as "empowerment" and "total quality management." At the same time, she tried to reshape the department, announcing she would delegate more authority to Energy's field offices. "People at headquarters were so accustomed to [having control] that the way the field offices worked was, 'The less you tell the bastards, the better,'" she recalled.[5]

Watkins watched the changes with skepticism. The admiral lamented that the department "is regressing to the very philosophy that created the mess we found in the late 1980s" by allowing "the old guard" to revert to its old ways by giving on-site contractors more control.[6] New Mexico's WIPP supporters, however, were pleased. They were already acquainted with O'Leary—her husband, John, had been New Mexico's natural resources secretary and later a top department official working on WIPP. In the early 1980s, it had been John O'Leary who angered New Mexico politicians with his apparent promise of a state veto. Before his death in 1987, the two O'Learys ran an energy consulting business in Washington. While traveling out west, they often socialized with Carlsbad's Louis Whitlock; as Energy secretary, O'Leary remained in close touch with him, continuing the top-level access the Carlsbad contingent had enjoyed with Watkins. "She understands the importance of WIPP to the nation's future," Whitlock said.[7]

Even as she listened to familiar voices, O'Leary brought in outsiders offering other perspectives. Among them was Dan Reicher, the Natural Resources Defense Council attorney who had joined in the successful 1991 litigation over the WIPP land withdrawal. Reicher became the new secretary's deputy chief of staff and chief environmental counsel. "He said to me, 'The Big Kahuna of your first year will be in waste management,'" O'Leary recalled.[8] Another new face was Thomas Grumbly, who took over for Leo Duffy as waste cleanup czar. Articulate and politically astute, the 43-year-old Grumbly had a resumé as varied as O'Leary's. He had worked on Capitol Hill, supervised meat and poultry inspectors at the Agriculture Department, taught a course in risk analysis at Harvard, and run a nonprofit organization developing solutions on hazardous waste disposal. Grumbly got an immediate introduction to the politics of WIPP. After being sworn into his new job, the first person he met with was Idaho governor Cecil Andrus, who pressed him to continue supporting tests there. "To not move forward with doing on-site testing would be like building a nuclear reactor, removing its core, taking it elsewhere for analysis, reinstalling it and then announcing immediately that it is ready for full fuel-loading and start-up," Andrus wrote in a subsequent letter.[9]

Grumbly, however, did not feel bound to the promises of previous administrations. To educate himself, he talked extensively with Reicher. "He had a lot of experience with WIPP and I didn't know anything about it, which was good, because I didn't come armed with any of his prejudices," Grumbly recalled. "That was one of the reasons I got the job in the first place. I knew about hazardous waste, but I didn't have any baggage in the nuclear waste debate. What I did bring was a certain history and maybe some skill in consensus building, and I tried to apply that to this area."[10] To build consensus, Grumbly convened a meeting of scientists, environmentalists, Energy officials, and Carlsbad residents. He suggested the department "may have put the cart before the horse" in moving waste underground.[11] Such candor was a far cry from Watkins's declaration that the project needed to be used whether or not Congress passed a bill allowing it.

Ending the Test Phase

Initially O'Leary and Grumbly cautioned they would go forward with tests only if they yielded "significant" scientific findings. Aware of the need for a backup plan if the EPA rejected the tests under the "directly relevant" criteria, the department announced in March that it would review the test program. In an abrupt departure from its past position, the initiative explored the idea of not bringing waste to the site.[12] The review dismayed many of the project's supporters, including lawmakers who had voted for the land withdrawal legislation a year earlier as a direct result of the department's insistence it needed to bring waste to meet the long-term standards. But New Mexico Democratic representative Bill Richardson, who had unsuccessfully fought for meeting the standards before any waste was brought, was jubilant: "I feel vindicated," he said.[13]

Such a change in plans reflected the growing reality that putting waste underground would be far more complex than had once been envisioned. In May, a council providing technical and policy advice to the EPA added to the complexity. It called for a prohibition on testing until the Energy Department could specify where it would send waste if it became necessary to retrieve it from the site. Officials at the EPA cautioned the National Advisory Council on Environmental Policy and Technology that requiring the department to name a site would be politically difficult. They said it could place them in the middle of a political battle between Energy and states wanting waste to go to New Mexico without it returning to them. But panelists agreed it was practical for the EPA to know a location. Failing to do so, New Mexico assistant attorney general Lindsay Lovejoy argued, would be "an exercise in unreality."[14]

Sandia National Laboratories scientists working at the site found more evidence that testing would be a complicated endeavor. They learned that many wastes could not be characterized well enough to obtain the approval to ship them across state lines. The result was that plans for the initial bin tests used "simple" waste forms, such as glass rings used in cleaning. The rings, however, generated little gas to be measured—the ostensible reason for conducting the tests. Separate tests called for adding brine to bins of waste, an approach that Westinghouse Electric Corporation officials operating the site opposed for safety reasons. The situation led to many internal arguments. "There's a world of difference between wanting to run an experimental facility, which is what we needed to do for the bin tests, and run an operating repository, which is what the Westinghouse [contractors] wanted to do in Carlsbad," Sandia scientist Al Lappin said. "The long and short of it is, they didn't want anything to do with brine underground because from an operational perspective, it gave them nothing but headaches. . . . If you have a brine spill underground, you've contaminated a potential nuclear repository. So, after a lot of thrashing around, the decision was made [to not] allow the addition of brine to any of these bin tests."[15]

To no one's surprise, the department's independent review of the experiments reached the same conclusion as that of the National Academy of Sciences a year earlier: "There is no scientific, regulatory or operational imperative" to perform the tests at WIPP.[16] New Mexico's Environmental Evaluation Group (EEG) formally called for abandoning the tests, pointing to the continued fragility of aging underground test rooms.[17] And the radioactive waste task force warned that an insistence on underground experiments had "the potential for undermining the department's position and eroding public trust and confidence."[18] To Grumbly, it was clear the future lay in a different direction. "We developed a strategy paper that basically laid out what we were trying to do," he recalled. "Fundamentally, it was, 'We're going to do a lot of real testing aboveground in the interest of trying to get the National Academy of Sciences and the people that they would influence in a consensus.'"[19]

It fell to Grumbly to inform O'Leary of his decision. Although she was disappointed—she saw opening the plant as a way of paying tribute to her late husband—she endorsed the approach as long as Grumbly sold it to the powerful pro-nuclear Senate Energy Committee chairman, J. Bennett Johnston of Louisiana. In traveling to Capitol Hill, Grumbly stressed the estimated cost savings—more than $100 million—from not having to ship waste to New Mexico, as well as the possibility of an accelerated disposal decision from the EPA. Faced with those facts, as well as the mounting scientific evidence against testing, Johnston acceded. So did New Mexico Republican senator Pete Domenici, the plant's other influential political patron. He said he supported the decision "because it puts the science first"—a view in line with the one Richardson had implored Congress to adopt a year and a half earlier.[20]

Having obtained the political backing it needed, the department announced in October 1993 that it would do tests at Los Alamos National Laboratory and other sites. "This is a major break with the last administration's approach, which frankly did not give full consideration to the concerns of the scientific community, EPA, and the public," O'Leary said in making the announcement.[21] The plant could no longer be considered a research and development project; it was now an operational disposal facility in waiting. The decision ended four years of scientific wrestling in which the department went from declaring that waste shipments had to begin in 1991 to admitting they now could wait until sometime around the turn of the century. In a sense, the emperor finally admitted he had no clothes.

In New Mexico, there was much relief. To environmentalists, the scientific evidence against testing had accumulated to the point where it forced the department's hand. "We screamed a lot about the test phase, but privately we knew there would never be a test phase," said the Southwest Research and Information Center's Don Hancock.[22] Sandia's Wendell Weart said the preparations for tests were not entirely wasted, as they enabled scientists to learn about the organic components of the waste material and other areas. But the process "clearly was a diversion," he acknowledged. "The people who spent a lot of time on that could have spent it doing other things."[23] The costs were significant: the department spent more than $20 million to design, manufacture, and buy equipment for the tests; $40 million to maintain the site and make safety preparations; and another $40 million to prepare and characterize waste at Idaho National Engineering Laboratory for shipment. "It almost seems scandalous that it took so long for DOE to reverse its decision," said the EEG's Lokesh Chaturvedi.[24]

The decision to end the test phase ranks as perhaps the most significant instance in WIPP's history when scientific considerations superseded political ones. A favorable political climate helped make it possible. The Cold War had ended, and nuclear weapons were not the hot-button issue of a decade earlier. Indeed, the government had ceased to conduct actual explosive tests. Partisan politics between Congress and the executive branch were low, as Democrats now controlled the White House as well as the House and Senate. Production of plutonium triggers at Rocky Flats—and waste shipments to Idaho—had stopped. At the state level, neither Andrus nor any other governor was willing to wage another storage battle. Politicians were unquestionably restless about opening WIPP but confident it would happen someday. In New Mexico, meanwhile, public opinion remained consistent among a majority of residents that the project was not ready. According to the University of New Mexico's Institute of Public Policy, the combined number of respondents who believed the plant either was unsafe and should never open or should only open after major changes remained around or above the 60 percent mark in the early 1990s.[25]

In retrospect, abandoning the idea of testing with radioactive waste probably enabled WIPP to move forward in much the same way that the Energy–New Mexico agreements of earlier years staved off protracted federal-state battles. The move certainly headed off litigation that could have led to delays and, in the eyes of some officials, shut down the project. "If we had pursued that test phase, that thing would be dead as a doornail right now," Grumbly said years later. "It would have given all the people who were against it just cause—really just cause—for saying that their rights and sensibilities were offended." He said the decision proved science need not always bow to politics: "The trick is to try to get the science working in the right way with the politics so that when you come down to the endgame, the scientific community is at least visible and supportive of what the right policy is."[26] It was a trick that the department had rarely bothered to learn.

Atomic Pork and Listening to "Stakeholders"

By this time, O'Leary had continued her commitment to changing the agency's culture. In May, she ordered the declassification of millions of documents as part of the new "openness" policy. Seven months later, she expressed revulsion and regret at the news that people had been injected with plutonium without their knowledge or consent in the late 1950s. The experiments had been known about for years but resurfaced when a Pulitzer Prize–winning series of articles in the *Albuquerque Tribune* put a human face on the issue by identifying five of the human test subjects. The secretary's reaction to the series generated global news coverage and led to the subsequent creation of a White House advisory committee. To O'Leary, such a controversial acknowledgment of a grievous past practice was intended to shore up her agency's credibility. "After all we had attempted to build in terms of public trust," she said, "there was no other way to go."[27]

If O'Leary took the lead on such headline-grabbing issues, she let others dictate policy in more mundane areas. She yielded in December 1993 to aggressive lobbying from Carlsbad and created a new project office in the city that combined the functions of the WIPP Project Integration Office in Albuquerque and the Project Site Office in Carlsbad. The new office oversaw not only the site but related activities at all out-of-state locations shipping transuranic materials to New Mexico. O'Leary and Grumbly contended that the new office was not a political pacifier to Carlsbad. But there was no denying that both were eager to do what they could to ensure that the community remained one of the few places on the planet willing to host a nuclear waste storage site.

The result was that five months after the tests' cancellation, the WIPP workforce was around 860 employees, the largest number ever. Spending on the project reached $185 million a year, higher than it had ever been. "I think the secretary went out of her way to help us," said a grateful Mayor Bob Forrest, whose family's tire store and motel benefited directly from WIPP's business.[28] But if Carlsbad residents were placated, others were not. Critics derided the added staffing and budget as "atomic pork." They received considerable ammunition when a department inspector general's audit concluded the project was overstaffed. "With the abandonment in plans to receive waste at WIPP for the test phase, it does not make sense that staff levels should remain relatively constant," the audit said. Energy officials disagreed with the findings but conducted their own investigation and decided to whittle down about 15 percent of the workforce through attrition.[29]

By this time, staffing decisions were made by the man selected to head the Carlsbad WIPP office: George Dials, a 45-year-old West Virginia native. Dials had master's degrees in nuclear engineering and political science from the Massachusetts Institute of Technology and had been a decorated infantry commander in Vietnam. Before coming to the plant, he had been an assistant manager at the department's Idaho field office. He was a forceful, no-nonsense manager in the mold of Watkins. "He was a can-do, take-that-hill person," Grumbly said. "When I saw his background, I thought he would be able to get us from A to B."[30] Dials described himself as having "a reputation in industry and management as someone with the ability to get things done."[31]

At WIPP, getting things done amounted to developing a compliance application that could withstand the EPA's scrutiny and the inevitable legal challenges. It also meant obtaining a Resource Conservation and Recovery Act (RCRA) permit from New Mexico's Environment Department

Figure 28: *Energy Department officials Thomas Grumbly (left) and George Dials (Department of Energy photo).*

for the mixed hazardous and radioactive waste that the state was given the authority to regulate. Attaining the first goal began in December 1993 when EPA officials issued final amended environmental and health standards for WIPP. The regulations were more stringent than the ones put forth in 1985 that were ruled inadequate by a federal appeals court. The acceptable releases of radioactivity to groundwater could not exceed levels allowed by the Federal Safe Drinking Water Act, while the 1985 standards permitted levels up to four times the drinking water standard.

Supporters grumbled about meeting regulations that were not in place at the time the facility was being planned and built. "That's the trouble with these waste-disposal projects," Weart said. "They're so long in evolving that regulators are forever passing new laws and applying them retroactively."[32] Environmentalists, however, remained unhappy with the standards. The standards lowered the annual acceptable radiation to the public from 25 millirems to 15 millirems—an amount roughly equivalent to two chest x rays. Although the EEG believed the 15-millirem level was sufficient, others wanted it lowered to four millirems. Criticism also came over several issues that were not directly addressed in the standards. One was the installation of man-made barriers at the site to act as an extra guard against radiation releases. Critics had called on the EPA to incorporate a barrier requirement into either the standards or the criteria being drawn up to enforce the standards. The standards also did not deal with the environmentalists' request to have the limits on releases from WIPP set as stringent as they were at high-level-waste disposal sites. One EPA official, Bill Gunter, told New Mexico officials that the standards did not touch on as many areas as some wanted because there was "no clear consensus on a number of things" and because of the tight deadline imposed by Congress in the 1992 land withdrawal law. "Time was a factor," Gunter acknowledged.[33]

Several months after the new standards were issued, Dials's office developed a "disposal decision plan" that identified key milestones and timelines for WIPP activities. Before such a plan was issued, Dials noted that there was "no regulatory framework, no body of laws or regulations that defined the process for opening the facility."[34] To the critics, such plans had an annoyingly familiar ring. Hancock charged that Dials was obsessed with schedules: "George is an old nuclear weapons guy. It's his way or no way."[35] Dials responded by accusing Hancock of blocking WIPP at any cost. "Don is ethically and professionally dishonest—he likes to tell half a story," he said.[36] The arguments reverberated back and forth in the news media, giving the impression that little had changed at Energy to engender trust.

The department did listen to complaints through its "stakeholders" effort, holding a series of formal meetings with a variety of interests. As much as the initiative marked a break from practices of the past, it did little to placate critics. The program "is essentially ceremonial," Lovejoy complained to O'Leary. "In the forthcoming stakeholder forum, [Dials's office] has attempted to assemble, on short notice, groups to discuss numerous aspects of WIPP—suggesting that the purpose is to let stakeholders sound off and be done with them."[37] Such comments illustrated the difficulty the department faced between trying to manage its own affairs and letting others tell it how to do things differently.

Meeting the 10,000-Year Requirement

Debates over internal management were far from the only problems confronting the Energy Department. On the technical side, it struggled with the single most difficult question about the project: How can its safety be proved for a period twice as long as the recorded history of mankind? The EPA had set the 10,000-year requirement for nuclear waste repositories based on what the agency described as a "world-wide scientific consensus" that such a period was a reasonable amount of time in which to reasonably predict geology, hydrology, and climate patterns. If WIPP could meet the standards for 10,000 years, officials concluded, it was likely to survive beyond then.

Not surprisingly, some critics challenged that assumption. University of New Mexico geologist and longtime WIPP opponent Roger Anderson disagreed that 10,000 years was long enough to ensure a site would be stable for a much longer period. Given the toxicity and longevity of plutonium, he suggested a more valid period would be 10 times the element's half-life, or 240,000 years.[38] Others in academia expressed misgivings about the overall lack of effort expended to address such a formidable requirement. "The nation needs to find answers to two essential questions: Can the geological formations be counted on to remain stable for 10 millennia? And can human beings be counted on not to interfere?" asked Yale University sociologist Kai Erikson. "A very considerable amount of research has been done on the first of these questions. . . . But the government has done very little research on the second question."[39]

The question of deterring future human interference was very much on the minds of Sandia scientists. With the project in an area rich with mineral resources, one of their main concerns was how to warn an oil or a gas company that might be interested in drilling underground near the waste. Ordinary "keep out" signs could not be guaranteed to last; for that matter, even the English language could not be assumed to survive 100 centuries. As environmentalists pointed out, the Old English of 1,000 years ago was not readily comprehended. They cited a line from "Beowulf": "Bealocwealm hafað fela feoheynna forð onsended." Even the translated meaning, "Death-qualm has sent many a folk forth on their way," was hard to discern.[40] Clearly such a sociological undertaking was beyond the grasp of scientists and engineers trained in provable technical matters. "This," Weart said, "is an area that borders on science fiction."[41]

The task of developing a WIPP warning for what amounted to an eternity fascinated the media. Articles in newspapers and magazines around the world about the issue brought home the monumental and unprecedented limits of scientific truth and certainty. Many did little to hide their skepticism. "Why should anyone assume that it is possible to create a structure that will outlive any previous empire from Mongolian and Ming to Roman and Russian?" asked one British newspaper. "It is not as if the meaning of Stonehenge, which is a mere 3,500 years old, is crystal clear to modern man."[42]

Recognizing their limitations, Sandia officials turned to outsiders. In 1990, the laboratory put together two panels of 16 professionals from various fields who considered ways to protect WIPP against future intruders. One group was charged with trying to predict what life might be like in the year A.D. 12,000. Another was asked to develop petroglyphs and warning markers that might withstand the test of time. The panelists included sociologists, climatologists, and historians. One of them was Gregory Benford, a physics professor at the University of California-Irvine and a well-known science fiction author. Benford was astonished when a Sandia official first called him to outline the task. "That's impossible, of course," he told the representative, who responded by pointing to the political need for scientific assurance: "Sure, I know that. But this is Congress."[43]

Benford offered a detailed account of the effort in his book *Deep Time: How Humanity Communicates Across Millennia*. After much discussion, the panelists believed two elements of future scenarios most directly affected the likelihood of inadvertent intrusion: political control of the region in which the site was situated and the pattern of future technological development. It envisioned the possibility of "miner moles" that would tunnel deep underground looking for neglected mineral deposits. Such a scenario seemed to mean that markers would have to be deployed that were magnetically discernible from above, beside, and below the repository. In fact, each of the expert panels endorsed the idea of large-scale marker systems that could offer redundant protection as some of their parts eroded or vanished over time. "Both panels thought along truly gargantuan lines," Benford wrote. "Even in the future, the old equation big = important will probably hold."[44] The panelists also agreed on the need for a "central chamber" for detailed messages, after the manner of the deep vaults in the pyramids of Egypt.

After Benford and the others fulfilled their responsibilities, Sandia brought in a second group of 13 more individuals to do the work of designing warning markers. The invitees again came from varied backgrounds, including linguists, artists, archaeologists, and even NASA scientists studying extraterrestrial life. To ward off future curiosity seekers, the second group initially considered not marking the WIPP site at all. But the group decided it was better to risk calling attention to the potential hazards than leaving them unexplained. One panelist unsuccessfully suggested opening up the project to artists in a sort of "radioactive Olympics." That drew an indignant response from Michael Brill, an architecture professor at the State University of New York at Buffalo: "I'd die before I let the art world come anywhere near this."

The panelists formed two separate teams. Brill's group developed several flamboyant proposals, including a field of 30- to 70-foot-high stone spikes and a "landscape of thorns" consisting of concrete blocks scattered over a square mile. The design that the panel recommended was the "menacing earthworks system," which would employ 1,000-foot-long, lightning-shaped earthen berms radiating outward from a flat and barren landscape. The second team, meanwhile, decided to appeal more to intellect than to emotion, believing that scary spikes and other similar designs would convey an ambiguous message. It envisioned a granite shelter containing a diagram of the periodic table of the elements that highlights radioactive materials, such as plutonium and uranium. Such a shelter would be complemented by time capsules bearing clay and glass tablets warning off intruders and a "millennial marker" that allowed future generations to employ astronomy to date it. Both panels, despite working separately, did reach many of the same conclusions. They agreed on the need to use materials with minimal value, lest they be salvaged, and that would be designed to minimize their exposure to the environment. They also concurred that written messages should be conveyed in different languages, including English, Spanish, Russian, and Mescalero Apache, spoken by the Indians who were native to the area.[45]

Sandia rejected relying heavily on archetypes in its final report to the EPA. It adopted a direct communication strategy incorporating the "information center" concept along with smaller monuments on the perimeter and outer boundaries of the site. Relatively small markers would be buried at randomly selected locations and depths. Skepticism continued from critics, who remained convinced that warning signs would not stop the effects of an inevitable radiation leak. Weart acknowledged that the conclusions on how to proceed were difficult to defend because "there is no expert better than I am on what's going to happen over 10,000 years."[46] But Benford believed the process marked a start toward coming to grips with an important issue. He noted that the WIPP markers "will be our society's largest conscious attempt to communicate across the abyss of deep time. There will be others, for the problem of virulent waste will not go away. . . . How we present ourselves in these ancient sepulchers may be our longest-lasting legacy. It is sobering to reflect that distant eras may know us mostly by our waste—and by our foresight."[47]

Congress Grows More Partisan

If dissent from Congress was kept to a relative minimum during O'Leary's decision to end the test phase, the Republican takeover of Congress in the November 1994 election returned politics to the front burner on Capitol Hill. Conservatives from states such as Idaho, Colorado, and South Carolina who clamored for waste to be shipped from their states were now the chairmen of committees and subcommittees and able to set the agenda. Many of them believed the EPA was unduly burdensome and insensitive to local concerns. The fact that the White House remained under Democratic control only heightened animosity toward the executive branch.

The Energy Department—and O'Leary herself—became caught in the political crosshairs. Republicans in the House pushed legislation to abolish the department and turn its weapons production and waste management responsibilities over to the Department of Defense. Because powerful lawmakers such as Domenici—who now controlled WIPP's purse strings as chairman of the Energy and Water Appropriations Subcommittee—opposed dismantling Energy, the proposal went nowhere.

O'Leary, however, came under prolonged attack. *The Wall Street Journal* reported in November 1995 that her office had paid a consultant $43,500 to identify and rank reporters who generated unflattering stories about her department. O'Leary defended the program as a legitimate effort to assess how well the agency was conveying its message, but Republicans howled with indignant outrage; even a senior White House official called the effort "monumentally dumb."[48]

Complaints about the secretary grew louder in the ensuing months, when news stories appeared noting O'Leary's 16 foreign trips during 1995, some of them aboard a luxury jet that had been used by pop superstar Madonna. On one weeklong visit to South Africa, O'Leary brought along 51 department employees plus an entourage of 68 executives, academics, and others, including a video crew. The trip cost taxpayers $560,000.[49] Once again, O'Leary defended such trips as "trade missions" and said the planes were selected on the basis of a competitive bidding process. But her excuses were drowned out as her image of boldness and candidness in disclosing the 1950s radiation experiments was forgotten. Instead she became a sort of poster child for government waste and abuse, with Republicans and newspaper editorials demanding her resignation. Although she stayed on, she was in a weakened political position.

When not attacking O'Leary and her agency, Republicans made clear their determination to bypass the bureaucratic processes set up for both commercial-high-level and defense wastes in the interest of political expediency. For commercial waste, they pushed legislation over the objections of the Clinton administration to establish a temporary aboveground storage site at Nevada's Yucca Mountain as studies continued on a permanent repository. They argued that the viability of the entire nuclear industry was at stake because the studies were taking too long. Nevada's congressional delegation, however, managed to stall the measure long enough to block final action before adjournment in 1996.[50]

As for WIPP, lawmakers from states with waste destined for the plant saw an opportunity to correct what they regarded as O'Leary's mistaken decision to cancel the test phase. In May 1995, Colorado's Dan Schaefer, chairman of the House Commerce Energy and Power Subcommittee, joined with Michael Crapo of Idaho and Joe Skeen of New Mexico to introduce a bill to amend the 1992 land withdrawal act by streamlining the regulatory process. Their bill included many changes sought by the department, including the elimination of a 180-day waiting period between the decision to operate WIPP and the start of waste acceptance underground. But the most controversial provision took the EPA out of the process altogether, leaving the Energy Department to determine the site's safety. "EPA," Skeen declared, "has no base of knowledge in this area." The legislation also called for shipments to start by March 1997, a year earlier than the department had projected. Predictably, environmentalists and EPA officials were furious with what they regarded as further congressional meddling. Even Dials argued against removing the EPA as the plant's regulator. "The department's history of self-regulation is, in many analysts' views, a chief reason for many of our waste and contamination problems," he warned.[51]

Three months after the Energy and Power Subcommittee passed the bill, Idaho Republican senator Larry Craig began pressing the Senate to follow suit. He introduced a measure that set a 1998 opening date and eliminated the 180-day waiting period. Unlike the House bill, though, Craig's legislation retained the EPA as WIPP's regulator. Craig attempted to stress the need for urgent action by pointing to an agreement negotiated by Idaho Republican governor Phil Batt—Andrus's successor—with the Energy Department and U.S. Navy. Batt, a more low-key politician than the flamboyant Andrus, struck the deal in response to continuing political pressure to move wastes out of Idaho. The agreement in essence said that in return for accepting 1,113 shipments of high-level spent fuel rods from navy warships and Energy projects, Idaho would not become a permanent dump. It called for most of the waste at Idaho National Engineering Laboratory to be cleaned up and shipped out of Idaho by 2035. If those goals were not met, Idaho could suspend the shipments at any time and fine the federal government $60,000 a day. The agreement also called for the Idaho lab's waste to be shipped to WIPP by April 30, 1999.

The decision ended a lawsuit in U.S. District Court over the department's environmental impact statement for fuel shipments to Idaho. In justifying the deal, Batt lamented that the defiant Andrus-style tactics of the past no longer would work because of the likelihood that the department could prevail in court. But Idaho environmentalists and others strongly disagreed. They argued that instead of limiting shipments to 40 years, the deal could cause the federal government to ship permanently into the state once it paid a fine as a cost of doing business. They also noted that unlike the language that New Mexico had gotten into the 1992 WIPP land withdrawal, the agreement contained no extra money for their state. Polls showed that 80 to 90 percent of respondents opposed the deal. Craig, for his part, saw his legislation as a way to help the embattled governor. "He was taking hits, and I was trying to offer some cover and do something here that demonstrated Congress's ability to move" on the waste issue, he said.[52] The bill also was intended to help Craig politically; his challenger in the 1996 Senate race, Walt Minnick, had made an issue of his commitment to removing nuclear waste from Idaho.

Recognizing the need to tone down the House bill, Schaefer got the Commerce Committee to approve an amended version in March 1996. It gave the EPA a greater role in certifying health and safety requirements but continued to draw criticism from environmentalists. With the full House now in position to consider the measure, the Senate moved to assert its dominance. Domenici and other senators announced in June they would offer an amendment to the fiscal 1997 defense authorization bill that would take many of the noncontroversial portions of the House bill and express the sense of Congress that WIPP should open by November 1997. Tacking the measure onto the defense bill was politically faster than following the normal legislative process.

Such an action was possible because the amendment had been toned down to the point that both the EPA and Energy could accept it. In addition to keeping the EPA as the regulator, language was added at the insistence of Democratic New Mexico senator Jeff Bingaman to clarify that WIPP should open by that date only if the EPA ruled that all health and safety standards were met. Bingaman's support of the amendment eliminated partisan conflict and cleared the way for its adoption in the defense bill, which Clinton signed into law. Bingaman decided not to oppose the bill after talking with EEG and Energy officials. "He asked me directly what I thought about it, and I said, 'As long as EPA stays in the process and gets to make the decisions, this is okay,'" Grumbly said.[53]

Environmentalists, however, remained deeply disappointed. Although the amendment continued to give the EPA the authority to determine compliance, it subjected the agency to what they regarded as too tight a time schedule. Among other things, the 180-day time frame that had been in place before shipments could start was shortened to 30 days. That one-month period, they argued, was not nearly long enough to give Congress and the public an opportunity for an adequate review. They also were upset the plant no longer had to comply with certain regulations on mixed radioactive and hazardous wastes that the Energy Department contended were unnecessary. This exemption eliminated the need for the department to receive the EPA's approval of its "no-migration petition" that had been granted several years earlier. In all, they complained, the amendment gutted the land withdrawal act.[54]

The critics were as upset with the process as they were with the result. Unlike the 1992 land withdrawal bill, no Senate hearings were held on the amendment. A frustrated New Mexico Attorney General Tom Udall lashed out at Domenici for being a "cheerleader" for the facility. "He doesn't want to see the kind of scientific scrutiny that the 1992 act mandated," Udall told a reporter. Domenici pointed to the backing of Bingaman "and 98 other senators."[55] Despite the objections of Udall and others, it was clear that no one in Congress saw much political mileage to be gained out of objecting to such a narrowly drawn measure over a facility in New Mexico that appeared to be well on its way to opening. In a period in which conservative Republicans were trying to overhaul bedrock environmental laws, environmentally minded lawmakers felt they had bigger fish to fry.

The EPA Encounters Opposition

If Udall and others were upset with Congress, they were equally dismayed by the Energy Department's approach to the certification process. In particular, they were unhappy with the guidelines issued for Energy to follow in demonstrating it could safely bury waste underground. The EPA document spelling out the so-called compliance criteria was completed in November 1995 but was not formally approved until three months later—after the Energy Department had been given a chance to review the criteria and lobby for changes, some of which the environmental agency agreed to. Officials at the EPA said the meetings were consistent with a federal Office of Management and Budget requirement that two federal agencies talk to each other before regulations are issued. But critics said the regulatory review interfered with the EPA's rule-making autonomy and shut the public out of having a chance to make their case. "What occurred was a subversion of the public process set out by Congress," Udall said. He filed a lawsuit in 1996 that ultimately was consolidated with two similar petitions filed by environmental groups and Texas's attorney general.[56]

The lawsuit exposed the limits of public confidence in bureaucratic decision making on nuclear waste disposal. In the eyes of WIPP's supporters, it was more than enough that Congress gave the EPA the power to be the independent regulator of WIPP. To WIPP's critics, however, the fact that the agency was part of the same administration as the Energy Department ensured it would lack the autonomy it needed. By this time, they had been conditioned to mistrust authority and to assume relevant information had been concealed from them. Some of the critics suggested that nothing less was needed than having the Nuclear Regulatory Commission (NRC) be redesignated as the plant's regulator. They said its authority as the civilian waste regulatory agency arguably placed it in a better position to oversee WIPP. But Congress remained opposed to merging the authority over military and commercial nuclear facilities. As a result, the other alternative that critics put forth was a new independent regulatory agency similar to the NRC. As they noted, though, such an approach had its disadvantages—an increased bureaucracy, added costs, and the time lost while the new agency mastered the regulatory task at hand.[57]

In addition to concerns over the certification process, critics continued to raise new questions about the repository's adequacy. They said lead-tainted brine seeping into the site from an aquifer about 75 feet below one of the four shafts leading into the site's storage rooms posed a safety hazard. The lead came from a shaft liner, drilling materials, and chain link fences that had been used in the site's construction that had flaked off and mixed with water. Energy officials discounted the idea that the seepage was a major problem.[58] Another hazard cited was the "Hartman scenario," named for southern New Mexico oilman Doyle Hartman. A Santa Fe jury ordered Texaco Corporation to pay Hartman $2.9 million in December 1994 after he argued that the company injected water into the ground at its well and that the underground water—moving through high pressures near WIPP—blew out his own well 20 miles from the site. Critics wondered if water flowing from other oil wells could reach WIPP and eventually carry waste to the surface.

Udall's office hired John Bredehoeft, a California hydrogeologist and former member of the National Academy's WIPP panel, to assess the implications. Bredehoeft concluded that under a worst-case scenario water could flood the repository and violate the EPA-established radiation release limits. Sandia, however, did its own calculating using a different set of assumptions and concluded that the chance of a flood was much less probable. "I wouldn't want to say it's impossible, but it's extremely unlikely," Weart said.[59]

The academy, for its part, was unambiguous in delivering its broad-ranging assessment of the plant's overall safety. It issued a report in October 1996 declaring that the repository "has the ability to isolate transuranic waste for more than 10,000 years" as long as "it is sealed effectively and remains undisturbed by human activity." It found that human exposure with radiation releases "is likely to be low compared to U.S. and international standards."[60] The academy's study was immediately questioned as suspect by Hancock and others but gave the department and its boosters the independent seal of approval they had long been seeking. In essence, they argued that any concerns the critics raised were more or less moot. "Because the scientific community has now confirmed that the WIPP site is safe, it is even more likely it will open on time so we can get nuclear waste out of our state," Idaho's Craig crowed.[61]

Later that month, the compliance certification application for WIPP was completed. It consisted of 21 volumes of material, plus 50,000 pages of reference material. The submittal came nine months after the EPA reviewed the draft application and told the department that it lacked the necessary detail for an appropriate and thorough review for technical adequacy. Exhausted Sandia scientists who worked on the effort were confident the actual application would withstand the EPA's scrutiny. "There is probably no single piece of similar-sized real estate on the planet that has been more closely studied and thoroughly characterized than the WIPP site," said Margaret Chu, a Sandia chemist.[62] Environmentalists remained unimpressed. "The fact that the Department of Energy has spent $2 billion and produced a lot of paper doesn't mean that the paper is worth much," Hancock scoffed.[63]

The application was given to the EPA almost four years to the day after President George Bush signed the land withdrawal bill into law. O'Leary was in her final months in office and, like Watkins, lamenting she would never get to cut the ribbon at the opening ceremony. "By the time I was about to go, I burned with a passion to prove to someone that the Energy Department had the will, the guts, the might, the intellectual capacity, the diplomacy, to open a waste repository," she said.[64] But the length of the process did not bother others. "My view when it comes to controversial facilities like this is, take your time," Reicher said. "We're talking thousands and thousands of years. Take your time with the process, with the science, with analysis. Part of what we added to this was, we did take more time, because there were all sorts of political pressures from various states, but there was really no rush."[65] Over the next four years, however, a fight with the state of New Mexico would lead critics to question whether there was indeed a rush—and would culminate in waste going to WIPP sooner than they felt was necessary.

8

"This Is Indeed Historic"

1997–2001: AFTER FIGHTING HIS HOME STATE, RICHARDSON OPENS WIPP

Bill Richardson sat stone-faced before a shining bank of television lights and a skeptical band of Republicans inside the Hart Senate Office Building's second-floor hearing room. In the years since his 1992 attempt to force the Waste Isolation Pilot Plant (WIPP) to meet all Environmental Protection Agency (EPA) standards before opening, Richardson had become a renowned foreign policy figure on behalf of his friend President Bill Clinton. He traveled the world as a freelance envoy negotiating the release of political prisoners held in Iraq, Burma, North Korea, and elsewhere. He quarreled with Iraq's Saddam Hussein and smoked cigars with Cuba's Fidel Castro. His successes made headlines around the world and earned him an appointment as U.S. ambassador to the United Nations, where he spent a year and a half engaging other diplomats on flare-ups in the Middle East and Asia. Now, however, it was July 1998, and Richardson was trying to become secretary of Energy by fielding a barrage of questions on the unglamorous subject of nuclear waste.

Members of the Senate Energy and Natural Resources Committee interrogated Richardson at his confirmation hearing about his decision to offer White House intern Monica Lewinsky a job at the UN. But they spent even more time questioning how strenuously he would work to open Nevada's Yucca Mountain to bury high-level spent fuel accumulating at commercial reactor sites in 35 states. Clinton's refusal to sign legislation allowing for temporary above-ground storage near Yucca Mountain had left them incensed. "This administration has not worked with Congress, will not work with Congress and has gagged its secretaries on the issue of nuclear waste," charged Idaho Republican Larry Craig.[1] Craig and other senators also wanted to know Richardson's plans for WIPP. The New Mexico facility had cleared the biggest of milestones when it received the EPA's certification two months earlier. Nevertheless, New Mexico's Environment Department had not issued an expected permit for the "mixed" radioactive and hazardous materials that the state was given the authority to regulate under the federal Resource Conservation and Recovery Act (RCRA).

Richardson assured the senators that he shared their frustration. Although he acknowledged his past reluctance to bring waste into New Mexico, he cited the EPA's certification as evidence it could be done safely. He also invoked the need for the agency to meet its goals of closing Colorado's Rocky Flats plant by 2006 as well as moving waste out of Idaho National Engineering and Environmental Laboratory to comply with the April 1999 deadline imposed by the Energy-Idaho agreement four years earlier. "I support the department's decision to open WIPP," he said. "It needs to open so that Rocky Flats and Idaho can keep on schedule."[2]

Richardson's promise marked the start of a series of events that pitted him and his legal advisers against his home state's environmental agency. The ambitious politician who once admitted he was "never a big fan of WIPP" ultimately chose not to live up to a promise to the state that the Energy Department had made and allowed the trucks to roll before the mixed-waste permit was issued. Consensus between the department and its critics was by now nonexistent as political pressure to open the plant reached its peak and both sides dug in their heels. The situation re-exposed the tensions over the role for a state in regulating a federal project and led to a new wave of litigation that put the judicial branch in charge of determining WIPP's fate. The department eventually prevailed in court, and the first shipments arrived to great fanfare. But the long-awaited event did little to quell the animosity between the state and federal government.

Richardson moved from the United Nations to the Energy Department during a period in which the agency was adrift. After Hazel O'Leary's political troubles prevented her from continuing into Clinton's second term, the president picked Federico Peña in early 1997 to succeed her.

Peña, a former mayor of Denver, had planned to leave government after four years as Transportation secretary. But with Housing and Urban Development secretary Henry Cisneros retiring under a cloud of scandal, Peña's departure would have left the cabinet with no Hispanics, so Peña reluctantly agreed to take the Energy job. When it came to WIPP, Peña spoke from the same script Richardson would later follow, vowing to senators it was "one of the department's highest-priority projects. . . . Efforts are under way to open the facility for disposal of transuranic waste in 1997."[3]

As Peña sought to pacify an impatient Congress, however, less optimistic Energy officials warned of delays. In February, they told a House committee that the certification process could take longer than anticipated and push the opening to February 1998.[4] Furious Republicans responded by flexing their political muscle, demonstrating again how politics could clash with the scientific decision-making process. A strongly worded letter was dispatched to Peña and EPA administrator Carol Browner, charging both agencies with "internal inefficiency, lack of clarity and poor policy decisions," singling out the EPA's stringent standards for WIPP. "The fact that EPA and DOE cannot efficiently handle a project of such national importance, in many ways the symbol of both agencies' environmental management abilities, makes the present delays difficult to understand at times," the letter said. Its 16 signers included several powerful lawmakers: Senate Energy chairman Frank Murkowski of Alaska, Senate Energy and Water Appropriations Subcommittee chairman Pete Domenici of New Mexico, and House Energy and Power Subcommittee chairman Dan Schaefer of Colorado.[5]

Domenici was unapologetic in turning up the heat on the executive branch. He had come to regret the number of safeguards included in the 1992 land withdrawal law. "The people who are anti-low-level radiation had too much political stroke in putting this package together from the beginning," he said. "Those who knew there was little risk did not exert the power they should have, including me. We have the most overregulated, overbuilt facility that anybody could dream up."[6] Critics, though, regarded the congressional tactic as outrageous. They took the opposite view, contending the EPA had made its standards weaker than necessary. Five days after the lawmakers sent their letter to Peña and Browner, New Mexico assistant attorney general Lindsay Lovejoy appeared before a three-judge U.S. Circuit Court of Appeals panel in Washington to argue that the EPA's criteria were too vague and that there was no way to quantify their effectiveness. His arguments left critics hopeful that the judicial branch could again prevail over the legislative and executive branches that had failed them. "If the case is decided on the law and on the facts, we'll win," Don Hancock of Albuquerque's Southwest Research and Information Center said afterward. "But we're also cognizant of the fact that it's very tough. Nine times out of ten, when you challenge a federal agency, you lose."[7]

Hancock's caution proved well-founded. Two months later, the panel of judges ruled that the EPA's compliance criteria could stand. They disagreed that the environmental agency was caving to pressure: "There has . . . been no general abdication to the discretion of DOE experts." In addition, in rejecting the argument that some criteria were not proposed with enough clarity for critics to adequately comment, the judges appeared willing to grant the EPA latitude by accepting its approach to making decisions that were less specific than critics demanded. They noted the agency "tried to avoid prescribing specific design choices or technical decisions so that EPA does not have the unintended effect of making the facility less safe, thus hoping to allow the scientists and technical experts administering the WIPP, presumably those most knowledgeable about the facility, freedom to make reasonable judgments. . . . In light of the complexity and uncertainty of planning for contingencies over the next 10,000 years, this seems quite reasonable."[8] The ruling seemed to presage the difficulty critics would encounter in challenging administrative decisions based on scientific uncertainty about a process guiding a facility that was without precedent.

The EPA Reviews, Fights with Energy

Officials at the EPA denied political considerations played any role in their deliberations on certifying WIPP. But they acknowledged that the tight time constraints Congress placed on them did have an impact. As a result, they sought to review the application without getting entangled in bickering with Energy. Nevertheless, suspicion lingered on both sides.

"It took time for them to trust us to the point where they would share all of the information we needed," said Frank Marcinowski, acting director of the environmental agency's radiation protection division.[9] Energy's Thomas Grumbly, meanwhile, felt the regulators treated his agency like an industrial polluter. "They still look at DOE as being not that much different from the W. R. Grace Company," he said.[10]

Given the difference in approaches and philosophies between the two bureaucracies, as well as the complexity and magnitude of the task at hand, it was hardly surprising that disagreements arose. A substantial part of the application derived from Sandia National Laboratories' computer modeling of how the repository was expected to behave in the centuries after it closed, based on forecasts of groundwater movement, climate changes, and other factors. The EPA wanted more details about the parameters used to determine the models. The result was that its staffers, working with Sandia, ended up reviewing not only the results of the modeling but the 20 years of information that went into developing the inputs of the models. "We didn't feel that some of the parameters chosen were appropriate," Marcinowski said. "There's a lot of uncertainty in all of this. We wanted to make sure that all possible ranges were captured, and they were less willing to do that for us. We just told them they had to do it."[11]

The EPA informed the department in December 1996 that its application was technically lacking. It was not until May 1997 that the agency had received enough material to judge it complete. Among the issues it paid attention to was the so-called Hartman scenario that arose out of environmentalists' concern that injected fluids could fracture rocks and cause brine to travel for miles. Although an initial analysis determined such an event was unlikely, officials performed a second analysis, in part because they realized the scenario was likely to form the basis of a legal claim. It confirmed that an occurrence was unlikely and that if it did happen, the result would be inconsequential. Marcinowski said other areas were closely scrutinized. "There were a lot of analyses to double-check what DOE had done that were probably unnecessary," he said.[12]

By October, the EPA was comfortable enough with the application to give it an initial passing grade. The agency issued a proposed rule that the plant complied with its disposal standards and started a public comment period. Among the conditions it imposed were that the department demonstrate in advance of shipments that it could accurately characterize the contents of the waste drums and seal up the site with reinforced concrete barriers to reduce the movement of hazardous gases and prevent the release of radiation. "None of the radionuclides that get out will have [a harmful] impact on health or the environment," predicted Richard Wilson, acting assistant administrator of the EPA's Office of Air and Radiation.[13] Sandia scientists were gratified. "It all came together quite comfortably in the sense that we weren't right at the boundary of their standard, we fell quite a ways below it, even when we responded to their request to adopt more conservative parameters," Wendell Weart said.[14]

In New Mexico, however, some wondered whether the EPA was as scrupulous as it claimed. The Environmental Evaluation Group (EEG) accused the agency of ignoring "serious technical questions": "The proposed rule has accepted the DOE viewpoint on most of the issues, sometimes without any questions, and others after minor clarifications," the group said in an analysis. The group also accused the regulators of being unresponsive to its comments: "The EPA reaction to our reviews and suggestions has been slow and apparently driven by legal considerations."[15] Hancock and Attorney General Tom Udall suspected the regulators were more interested in meeting a deadline than in doing a substantive review. They accused the agency of ignoring possibilities that could cause massive radioactive releases and violate the standards, as well as dramatically underestimating the likelihood that drilling activity could hit a pressurized brine reservoir. Clearly, Congress's intent of making the EPA the arbiter of debates over scientific uncertainty was far from enough to instill public confidence in everyone.

Such arguments, however, had become difficult for critics to win. It was no longer enough to accuse the federal government of disregarding safety; instead, the debate had shifted to just how far it should go in the name of assessing it. In such a highly subjective and technical process, the EPA and Sandia could—and did—argue that their models and other efforts factored in all relevant scenarios. In addition, the critics lacked the scientific resources to put them on an even footing with the two agencies. Udall acknowledged as much in discussing his litigation over the EPA's compliance criteria before the ruling was issued. "A lot of the lawsuit up to this point has been how they handled the procedure and process issues," he said. "It's different when you get into second-guessing the health and safety issues that EPA has to rule on. We don't have the ability to go out and spend half a million dollars on scientists and second-guess the scientists."[16] Although the Attorney General's Office did turn to experts such as hydrogeologist and former National Academy of Sciences panelist John Bredehoeft to support its claims, it often depended on the EEG's analysis. The state watchdog agency, however, found no serious "showstoppers." Senior EEG scientist James Channell summed up his organization's position when the EPA issued its final five-year certification in May 1998. "It's a reasonable decision," Channell said. "You can't wait until everything is perfect, because if you did, you would never start anything."[17]

The certification was the biggest technical milestone WIPP had cleared. The Energy Department had finally received the independent blessing that observers said was needed to protect health and safety as well as promote public confidence. Despite the battling with Energy, most EPA officials considered the process successful. In an internal review, they concluded several steps needed to be taken in planning for the next certification in 2003, including making clearer to the department what the new application should contain. A separate review by an EPA-hired outside consultant also praised the agency for exceeding its requirements for public outreach. However, it said, "While keeping the public well informed about its actions, EPA was unable to involve the public in the key aspects of the decision-making in which they were most interested." The consultant found that the EPA could not share the results of its technical evaluations and that the agency's role was narrow in competing against broader public interests. "In many instances, EPA and the public were simply not addressing the same problem," it said. "EPA faced the problem 'how to make WIPP safe' while the public was addressing 'how to open/prevent opening WIPP.'"[18]

The bottom line was that the certification did nothing to stop critics from fighting WIPP. Two months after the EPA issued the ruling, Udall's office joined three environmental groups in filing a lawsuit with the U.S. Court of Appeals. The suit alleged violations of the notice-and-comment rule making and substantive technical errors in the certification. In essence, it argued that the EPA used the wrong procedures and ignored many of the comments it received. "I'm going to take every legal avenue I can to get the Department of Energy's attention," Udall said.[19]

Others in New Mexico government, however, were less inclined to go along. A new administration had come into power with the 1994 election of Republican governor Gary Johnson, a political novice who thwarted Democrat Bruce King's bid for another term. The constant opposition of environmentalists was again in conflict with the on-again, off-again backing of state government. Jennifer Salisbury, Johnson's secretary of the Energy, Minerals and Natural Resources Department, praised WIPP's "solid scientific foundation" and told EPA officials at the January hearing: "From a taxpayer's perspective, it simply makes no sense to abandon a $2 billion investment."[20]

The support from Johnson's administration coincided with a shift in New Mexicans' attitudes toward WIPP. Between fall 1997 and spring 1998, the percentage of residents surveyed by the University of New Mexico's Institute for Public Policy who favored opening the facility rose from 45 percent to 49 percent. At the same time, the corresponding number of residents who did not favor opening WIPP dipped from 50 percent to nearly 46 percent. The center suggested the change appeared to be influenced to the degree of progress being made toward opening: "As long as WIPP was held up in court or by regulatory agencies, most New Mexicans concluded that WIPP was not ready to open. Once the legal hurdles had been cleared, the weight of opinion shifted toward accepting that WIPP was ready to open."[21]

New Mexico–Energy Department Clashes

Despite Johnson's acceptance of the project, the governor did nothing to interfere with the New Mexico Environment Department's rigorous scrutiny of the mixed-waste-permit application. Energy officials had always anticipated the Resource Conservation and Recovery Act (RCRA) process would parallel the EPA review, with the awarding of the state permit occurring around the same time as the final certification decision. That assumption proved to be naive and created tensions between the state's wish to maintain some control over the project and the federal government's desire to serve other states by starting shipments. It was a classic—and unusually contentious—federal-state conflict.

New Mexico's regulation of hazardous waste at WIPP had extended back for almost a decade. In an example of how federal law provides states with an enforcement mechanism over mixed waste, state programs are authorized under RCRA to operate in lieu of the EPA. At the same time, the 1992 land withdrawal law required the Energy Department to obtain a state permit before the management, storage, or disposal of any waste. The state Environment Department asked Energy in August 1990 to submit an application in two parts: a "part A" version providing general information about the facility and a more detailed "part B" document. Energy officials submitted both applications in 1991 in order to bring in waste for the five-year test phase. After ordering two revisions, the state issued a draft permit in August 1993. But after the test phase was canceled that fall, the department demanded that a revised application be submitted.

By May 1995, Energy officials complied, submitting a 13,000-page part B application. Despite its length, the state faulted it for lacking important details. Energy officials responded by submitting more than 20,000 pages of extra material. Nevertheless, state officials found they were getting little of what they sought. They eventually concluded that continued requests "were unlikely to obtain additional information" and began developing a permit.[22] The Energy Department, for its part, had no apologies for taking so long. "This was a complex permit," said Mary Anne Sullivan, the department's tough-minded general counsel. "It is in the nature of these processes that you have an iterative submission."[23]

The Energy Department's Task Force on Radioactive Waste Management, in its 1993 study of building confidence in waste disposal programs, had anticipated such interagency conflicts over the RCRA permit. The study pointed to the lack of specific knowledge about the types and quantities of waste to be sent to the plant. "Because the cost and schedule implications for WIPP of having to develop a detailed description [of the wastes] are significant, the possibility of substantial disagreements arising with the state and federal regulators cannot be ruled out," the task force said. "Such a circumstance would hardly strengthen public trust and confidence."[24]

Before the EPA certification was issued, Energy officials in Washington left the RCRA matter largely to the Carlsbad office and its manager, George

Dials. In retrospect, Grumbly regretted not having been more closely involved. "I really had not envisioned it would end up being the ultimate sticking point in the equation," he acknowledged. "We relied on the people in the state who were strong supporters to be able to take care of that particular problem, and we didn't work hard enough."[25] As it was, continued pressure from pro-WIPP state lawmakers left the Environment Department feeling besieged. Its attorney, Susan McMichael, complained about the political-scientific tension: "Political pressure placed upon [the state] to expedite this process . . . serves no useful purpose at all except to further hinder this complicated process."[26]

Relations between the department and the state continued to erode over the permit application schedule. To McMichael and other New Mexico regulators, the complexity of the application made delays unavoidable. With WIPP moving toward meeting other technical requirements, they often wondered if the federal agency would try to bypass them. Initially Dials promised that would not happen. In February 1994, he wrote that his department "has no plans or intentions of disposing of any wastes (neither hazardous, radioactive nor mixed) in the WIPP prior to receipt of a RCRA Part B Disposal Phase permit."[27] But as pressure to open the plant continued, that pledge was abandoned. In a major policy change, the department proposed to ship nonmixed wastes that were only radioactive without containing any hazardous chemicals and thus not subject to state regulation without a state permit.

Two developments precipitated such a shift. One was a change in the 1996 law passed by Congress intended to speed up the project's opening. It exempted the department from the treatment standards and land disposal restrictions under RCRA—the thinking among lawmakers being that other parts of the act would still apply. They reasoned that because the department was going to have to prove the plant could contain radioactivity for 10,000 years, it made little sense to require it to do any extra work to show it could contain chemicals for a few decades. In addition, New Mexico Environment secretary Mark Weidler said at an October 1997 conference in Carlsbad that the department could dispose of nonmixed radioactive waste without a permit. Weidler added, however, that the federal agency still could not store or dispose of any mixed waste without the state's permission.[28]

To Dials, Weidler's acknowledgment paved the way for allowing shipments without state approval. "It's irresponsible to think we would respond any other way," he said in November 1997. But in pursuing such a move, the department gave the perception of flip-flopping in the interests of political expediency. In the final environmental impact statement for WIPP, the agency said the site would be eligible to receive waste "only after several additional conditions are met," including the receipt of the permit.[29] Other Energy officials defended the change in direction, arguing that Dials's 1994 promise was not legally binding. But the move damaged their credibility by making it appear they were more interested in opening the site than in following the law. "One of the things I have learned and would press my clients to take care to avoid is the kind of representation George Dials made," Sullivan said.[30]

To resolve the issue, Sullivan ended up copying what critics had done: going to court. In May, her office asked U.S. District Judge John Garrett Penn whether the 1992 permanent injunction he issued blocking the plant's opening remained in effect. The department contended the injunction was moot because of the subsequent congressional land withdrawal and the EPA certification. Udall, however, argued the court order remained in effect until the judge lifted it. His office joined four environmental groups in filing a motion seeking a preliminary injunction to prevent waste from going to WIPP. Penn responded by issuing an order setting a schedule to address the critics' motion.

A central issue in the case was whether WIPP was eligible for "interim status," a provision under RCRA for hazardous waste treatment facilities that were in existence before the law went into effect in 1980 or had received a "statutory or regulatory change" subjecting such facilities to RCRA's permit requirements. Interim status allowed facilities to operate without a permit during the application process. Initially Penn ruled that the plant did not have such status. But the U.S. Court of Appeals, in its subsequent July 1992 decision affirming the permanent injunction, reversed Penn's decision, citing an interpretation by the EPA that radioactive mixed wastes were regulated by RCRA in 1986. "Because RCRA does not define the type of change that qualifies as a 'regulatory change' under the statute, we defer to EPA's reasonable interpretation," the appeals court wrote. The judges did not, however, address the precise date of when WIPP itself received such a regulatory change, leaving the matter to Penn.[31]

As they litigated the issue, Sullivan and other department officials worked with the Environment Department over a controversy involving 116 drums of waste to be shipped from Los Alamos National Laboratory. The state determined in June that the department failed to adequately characterize the drums to prove they contained nonmixed radioactive waste, touching off a dispute. Shortly thereafter, the two sides reached agreement on a schedule and process for the waste, and in December, the department determined the drums were nonmixed. Agency officials said it demonstrated their willingness to work with state regulators. "We could have said, 'We're not working with you, and if you don't think we have interim status, sue us,'" said Energy attorney Paul Detwiler. "But we went through an incredible characterization effort where we spent a lot of money and a lot of people's time."[32]

The time it took to settle such issues prevented the plant from opening in 1998. But other unresolved disputes would set the stage for a final battle involving a judge who had already halted WIPP's opening once and a politician who had found himself on the opposite side of the environmentalists he had once supported.

Richardson Lets the Trucks Roll

Richardson arrived at the Energy Department in August, three months after New Mexico officials issued a draft RCRA permit and Sullivan informed them that nonmixed waste would be sent without a final state action. The former congressman was not the only influential policy maker thrust into the WIPP debate; New Mexico's Environment Department also had a new secretary. After Weidler was killed in a car crash in July, Johnson brought in Peter Maggiore to head the agency. Maggiore, a geologist like Weidler, had worked as its director of environmental protection until moving to the state Economic Development Department. At 41, he was a knowledgeable bureaucrat who sought to handle the permit in a straightforward manner. "I thought, 'Who's going to argue against good science?'" he recalled.[33]

Maggiore tried to persuade Richardson and Sullivan they should not ship nonmixed waste without a final RCRA permit, citing Dials's 1994 promise and statement in the application that all waste at WIPP would be handled as mixed. He warned that doing otherwise could lead to further delays in approving the permit by forcing him to reallocate staffers to review and analyze nonmixed waste destined for shipment. Similarly, he and McMichael stressed that the permit represented the state's only hands-on external regulation of WIPP; without it, a variety of requirements would not be met, from monitoring of groundwater to guarding against fires or explosions from hazardous wastes.[34]

As 1999 began and the court hearing before Penn drew nearer, the two sides made a stab at negotiating a settlement. The Energy Department began working with new attorney general Patricia Madrid, a Democrat who had been elected to succeed Udall. But Johnson's chief of staff, Lou Gallegos, became dismayed when he learned how little the department appeared to care about giving the state some control. The Environment Department "was never shown a settlement proposal from DOE that appeared to be in the best interests of the state of New Mexico," Gallegos complained in a letter to Gary Falle, Richardson's chief of staff.[35] Falle shot back a response contending the Energy Department felt "it was seriously misled" by the state's handling of the Los Alamos waste issue, because it had suggested in September that the drums might in fact qualify for disposal at WIPP. He noted that the state's permit might not become effective until January 2000. "Given the severe backlog of transuranic waste building up around the DOE complex as a result of the RCRA permitting delays, and our regulatory and other commitments to meet certain milestones for disposal of transuranic waste at WIPP over the next several years, beginning shipment of Los Alamos waste is vitally important," he concluded.[36]

As the bureaucrats continued to fight, environmentalists were given a final forum for their objections. The state held a lengthy series of hearings on the revised draft permit at which they hammered away at the department and its contractors for its lack of attention to detail and its alleged misstatements on technical issues. The hearings provided them with a chance to renew concerns that had been raised during the EPA permitting process, such as the lack of adequate characterization of wastes and the Hartman scenario. "I think it's possible to get a similar situation . . . in the vicinity of WIPP, and should that occur, it's potentially possible that a lot of fluid could move to WIPP," Bredehoeft testified at one of the hearings.[37]

Politicians, meanwhile, continued to throw their weight around. Domenici warned the New Mexico legislature that Congress could try to cut WIPP-related payments to the state if the plant did not open. "Pretty soon, maybe this year, somebody will raise the issue, 'Why fund it?'" he said.[38] In fact, Craig talked with Domenici about legislation that would strip New Mexico of its mixed waste permitting authority but decided it would appear too heavy-handed.[39] In Idaho, new Republican governor Dirk Kempthorne reiterated his state's pledge to hold the Energy Department to the April 30 deadline to move waste out of his state.[40] Richardson, for his part, described himself as "increasingly frustrated" by New Mexico's position. He told the Senate Energy Committee that "the goalposts . . . seem to be moving back."[41]

The highly charged atmosphere was reminiscent of 1992, when then Energy secretary James Watkins had desperately sought to open WIPP. Once again, the department was seeking to ship waste without obtaining an approval it had promised to seek. And once again, it fell to Penn to cut through the arguments on both sides and illustrate the power of the courts in determining the future of a nuclear waste project.

This time, though, the judge would not hold things up. On March 22, ten days after a hearing, he issued an order stating that his 1992 ruling did not prevent the shipment of the Los Alamos nonmixed wastes, saying the earlier injunction he granted "addressed the matter before the court at that time and nothing more." Penn went on to rule that the plant did have interim status, citing an interpretation by the EPA that New Mexico was given authority to subject the facility to RCRA's permit requirements in July 1990 and that the EPA's position "is a reasonable one." As evidence, he noted that was the date the state itself had recognized until after the department had filed its RCRA application. In addition, Penn said the state and environmental groups had not demonstrated "they will suffer irreparable injury" if the shipments were sent before a state permit was granted, nor had they had shown "a likelihood of success" in blocking the shipments altogether.[42] After a quarter century of controversy, WIPP had finally received the last official go-ahead—not

Figur 29: Albuquerque Journal *cartoonist John Trever's humorous take on the long-delayed start of shipments to WIPP (courtesy John Trever).*

from an Energy secretary or Congress, but a judge.

Penn's decision drove home the reality that critics no longer could rely on the courts. As long as the Energy Department maintained its commitment to using the facility and as long as it was willing to follow the guidelines set out for it to some debatable degree, the legal system apparently would look at its efforts with a benevolent eye. Richardson immediately announced that nonmixed waste would be shipped from Los Alamos. "This is indeed historic—for DOE and the nation," he said.[43]

Had the Environment Department boasted a stronger advocate for its case in the governor's office, it might have continued the fight. But Johnson promised not to interfere, calling Penn's order "really great news." Despite his preference for wanting to wait for a state permit, the governor was willing to defer to Richardson. "DOE is fully aware of the potential risks relating to that permit," he said, "and is apparently willing to take those risks."[44] Madrid was similarly deferential. "This is from a judge who was once sympathetic in ordering an injunction," she said. "I think that's a good reason not to appeal."[45]

At Energy headquarters, Richardson's decision to start the shipments before receiving the permit was viewed as necessary to prod the state to complete its permitting process. "Had we not pressed forward, we might still be waiting for a permit," Detwiler said in an interview more than a year later.[46] Richardson was not worried about any potential backlash in his home state. "As an issue, WIPP faded after the EPA [certification], and I am convinced it's not a very important political issue or environmental issue in New Mexico," he said in 1999. "I admire Hancock's persistence, but his objective has always been to shut it down."[47]

The move drew support from policy makers who had been in Richardson's shoes. O'Leary called the action "extremely courageous" and said she probably would have done the same thing.[48] Former waste management czar Leo Duffy said such an action should have been tried eight years earlier, after the appeals court's interim status ruling. "We could have shipped then and saved everything it's cost since then—a billion dollars or so," he said.[49] But some state officials outside New Mexico reacted with suspicion, wondering if it would set an unwanted precedent in their dealings with Energy on nuclear waste. "We often point to New Mexico as a reason not to sign any agreement with the federal government," said Robert Loux, director of Nevada's nuclear projects office. "It's clear they won't make good on it."[50] Not surprisingly, Hancock also was troubled. "WIPP demonstrates that if you give the feds enough power, the state gets screwed," he said.[51]

Without help from New Mexico's attorney general, it fell to Hancock and other environmentalists to mount last-minute legal fights in the days prior to the first shipment. All, however, were turned aside, and after being delayed by heavy fog, the first TRUPACT truck with the Los Alamos waste rolled up to WIPP's gates. The *Carlsbad Current-Argus* published a 12-page special edition filled with photographs and quotes from excited locals. Three weeks later, at a formal ribbon-cutting ceremony, Richardson smiled for the cameras along with the other dignitaries in attendance who had spent the past two decades wondering if such a day would ever come to pass: Domenici, Skeen, Senator Jeff Bingaman, Robert Neill, and Wendell Weart. "Having worked 25 years to get to this position, I had to feel a tremendous gratification," Weart recalled. "All of my work had come to some real, concrete fruition."[52] Quipped Bingaman: "I don't know what all of us are going to do with all the spare time now that WIPP's finally opened."[53]

Within two weeks of the opening ceremony, WIPP got its first out-of-state shipment. Forty-two drums of nonmixed waste from Idaho National Engineering and Environmental Laboratory joined the Los Alamos shipments underground just before Idaho's April 30 deadline. Former Idaho governor Cecil Andrus, who had crusaded for years for waste to be removed from his state, was not completely satisfied. He referred to the thousands of drums buried at the lab before 1970 that would not be buried at WIPP because they needed to be excavated. "It was symbolic and historic," Andrus said of the first Idaho shipment, "but there's a lot more to do."[54] Three weeks after that, 26 drums of nonmixed waste would leave Rocky Flats for the plant to a similarly enthusiastic reception.

By this time, the New Mexico Attorney General's Office had dropped out of the legal picture. Madrid announced in May that her office would withdraw from the lawsuit that had been filed under Udall's watch challenging the EPA's certification. She explained that the state had little chance of winning now that the process had moved from the legislative into the administrative arena. "It is extremely difficult to convince a court to overturn an administrative agency's discretionary decision making," she said. Environmentalists were livid, particularly since the announcement came on the eve of arguments before the U.S. Court of Appeals for the District of Columbia. The matter was dismissed a month later, after 14 shipments had been made to the site.[55]

Despite the setbacks in court, WIPP's critics refused to stop fighting. They continued to file lawsuits and hold public protests. Albuquerque activist Charles Hyder, now 69, even fasted for 82 days after the first shipment arrived.[56] The critics believed their arguments were bolstered by a series of flaws besetting the early shipments. In June, one of the shipping containers registered higher-than-normal levels of radioactivity, a problem that project officials attributed to the truck driving through an airborne by-product of naturally occurring radon. Around the same time, a plug was found missing on one of the containers, the waste-handling elevator

Figure 30: The initial waste drums shipped to WIPP were stacked and buried underground (Department of Energy photo).

was out of commission for weeks until a burned-out transformer could be replaced, and an audit of Idaho found problems in record keeping and waste-handling procedures, forcing a temporary suspension of shipments. "This is not a performance that inspires confidence or should inspire confidence," Hancock said.[57]

Fights Continue After Opening

New Mexico's subsequent issuance of a RCRA permit in October 1999 did little to end the conflicts between the federal and state governments. In fact, the acrimony only intensified. The Energy Department was upset over a number of permit provisions, including language that would force the agency to open more drums for inspection than it believed necessary. Sullivan contended such a measure would require additional visual inspections of waste containers and, as a result, increase radiation exposure to workers. She also opposed the state's desire to force Idaho to spend $7 million to reexamine waste that had already been judged ready for shipment.[58]

The biggest area of contention, though, came over another requirement that would touch off battles in the courts and in Congress. The Environment

Department required Westinghouse Electric Corporation, Energy's WIPP contractor, to provide a $110 million "financial assurance" to cover estimated costs and potential pollution problems associated with the eventual closing of the site. State officials said the language was similar to assurances imposed on the governments at hazardous waste landfills around the country. They noted that the state could not rely on Energy's promise that it would close the site safely. "The federal government routinely walks away from contaminated sites," said Nathan Wade, a spokesman for the state agency.[59]

Energy officials said the state had no business imposing the requirement. They noted that it called for the money to be placed in a private bank, but the Treasury Department declared it had no mechanism to set up such an account outside the U.S. Treasury. The officials said they also could not persuade Westinghouse to agree to a guarantee so far into the future beyond the company's five-year contract to operate WIPP. Richardson announced he would be forced to withhold $20 million in federal highway money mandated under the 1992 land withdrawal law to comply with the financial assistance clause and other requirements. The plant's defenders in Congress also cried foul. "Why we should pay taxpayer dollars to guarantee that we, the government, will do what we say we'll do when it comes time to close this facility is a waste of taxpayers' money," Bingaman said.[60]

Department officials were troubled that the New Mexico financial assurance requirement would lead other states to follow suit. "The reason we're worried about the precedent is because it has a cost," Sullivan said.[61] Within a few days, she went back to where other waste disputes had been settled: the courtroom. The agency filed suit over the permit in U.S. District Court in Albuquerque, a move that surprised Maggiore. At the time he received word of the legal action, he was meeting with Energy officials to discuss a resolution of the financial assurance issue. "I was a little taken aback with their aggressiveness," Maggiore said of the lawsuit.[62] Later, the department filed suit in the New Mexico Court of Appeals, where the state claimed that the dispute belonged. U.S. District judge E. L. Mechem, however, denied the state's motion to dismiss the lawsuit, ruling that the department's claims "clearly arise under and require determination of federal law."[63]

Domenici, meanwhile, pursued a political resolution. He got Congress to pass legislation that restricted the state from requiring financial assurances for storage. The state, however, continued to argue its right to impose an assurance requirement. Seven months later, Domenici added more legal language to prohibit the state from moving ahead. After Clinton signed that bill into law, Maggiore—by now under pressure not to hold up federal payments—relented and withdrew the requirement from the state's permit. The state also agreed to alter four waste characterization requirements in the permit that the department objected to in the litigation.[64]

Conflicts between the state and federal governments continued to flare into the waning months of the Clinton administration. In 2000, the Energy Department discussed several proposed modifications to its RCRA permit that critics immediately regarded as a significant change in WIPP's mission. The department sought approval in July to ask that the plant's surface storage facility be expanded to include a centralized waste characterization facility and allow waste drums to be opened on site—something project officials had testified at earlier hearings that they did not intend to do. They said the cumbersome system of analyzing the contents of waste shipments could be accomplished more readily at the plant, but environmentalists objected, saying materials that were prohibited from being shipped could end up coming there. They mobilized opposition to the proposal, and in September, the department withdrew the permit modification request.[65]

The department submitted a new request in June 2001 for a central characterization facility. Officials denied that what they sought amounted to an expansion of the plant's original purpose.[66] Three months after submitting its new request—and several weeks after 352 shipments had come to WIPP from 24 different sites, filling the first underground storage room to capacity—the Environment Department denied the Energy Department permission to modify the way it sampled what was inside the waste drums. Under the type of modification that Energy sought, it was allowed to implement the changes before the state approved them, even though it ran the risk of being rejected. The Environment Department, however, said the changes were more substantive and that the public should have a chance to submit comments before it made a decision.[67]

Stresses of State-Federal Interactions

The continually contentious situation involving the RCRA permit exposed the problems of giving a state limited—as opposed to total—enforcement power over the waste management decision-making process. If the state's limited authority conflicted with the federal government's and if neither side was able to reach a consensus, it would inevitably fall to Congress and/or the courts to resolve the impasse.

To Domenici, the conflicts indicated the shortcomings of the hazardous-waste law giving states authority in lieu of the EPA. "I wish I would not have put in the state's role under RCRA," he said in 2000. "It's not because I want to deny my state regulatory authority. But when you make that regulatory authority so narrow—and you have waste that is so much more dangerous being approved by EPA in a way that the statute says the state couldn't do anything about it—you're almost saying that you're inviting the state to be a player, but you're not."[68]

Other participants in the process, however, said that the New Mexico–Energy battle did not require taking the state out of the equation. Despite the many delays, political pressure, litigation, and overall contentiousness between the two sides, both Maggiore and Sullivan said the permitting process demonstrated the protracted give-and-take that should be present in regulating a facility as formidable as a nuclear waste repository. Although Maggiore suggested that RCRA could use some modifications to clarify the state's input into decision making, he said, "I'm not as pessimistic in terms of the process not working well. Yeah, it took too long, and yeah, it cost a lot, but it was the first of its kind. Having said that, I don't think it worked all that badly."[69]

Sullivan echoed Maggiore: "This is the world's first deep geologic repository for radioactive waste. As a matter of public policy, the process shouldn't be easy. Nobody should expect it to be smooth. And nobody should expect it to be sweetness and light between the operator of the facility and the regulators."[70]

Epilogue

In the beginning, supporters of the Waste Isolation Pilot Plant (WIPP) figured that opening the facility would be a relatively simple task. Technology had shown it was possible to build atomic bombs and walk on the moon; all they needed to do was keep a limited amount of radioactive material from reaching the environment in an area far from where anyone lived. "We didn't see any insurmountable hurdles at all," recalled Joseph McGough, the Energy Department's project manager when construction began in 1981. "It was a straightforward construction project."[1]

But the process that led to WIPP's use nearly two decades later was anything but straightforward. WIPP began during the Cold War but was developed in a post–Cold War world. Unlike other Cold War sites built in an era of secrecy, it had to navigate a complex bureaucratic maze in which safety was ill defined and based on considerable projection and scientific guesswork. It had to deal with a lack of continuity between changing presidential and state government administrations. And it was subject to unrelenting political pressures. During the Cold War, the federal government could act unilaterally in the name of national security, but the Energy Department of the 1980s and 1990s had to accommodate a host of skeptical outside interests clamoring for a say in what should be done. When those interests failed to get their way, they headed to court, an institution that became willing to approve major shifts in direction.

The department never anticipated such issues would be so problematic. For all of its attention to technology, it was slow to develop equally sophisticated and credible methods of dealing with the public. Such a fault was by no means limited to WIPP; a 1978 Nuclear Regulatory Commission (NRC) study acknowledged, "The persistent faith in a technological fix has produced a myopic vision of the waste management problem."[2] With WIPP, though, the expectations were higher for a culture that had been accustomed to doing things its own way and believed it knew what was best. When it was shown that the agency did not always know what was best at WIPP or other nuclear sites—or when there was no clear-cut way to determine what was best (such as conducting experiments with waste before meeting all standards)—its credibility was questioned. Disputes and delays followed.

The department's attempts in the 1990s and afterward to demonstrate the culture had changed proved difficult. Social psychologists have found trust in government institutions is a quality that is quickly lost and slowly regained; once an untrustworthy act is committed, subsequent honest actions do little to help. The situation has been a particular handicap in an endeavor as dangerous as storing nuclear waste, a risk many judge to be inherently unacceptable.[3] It hardly helped that nothing like WIPP was around; there were no precedents or benchmarks to follow. In the end, the department was fortunate in that critics were never able to make WIPP a national issue and muster enough popular or political support to kill it. Instead, the project developed its own momentum: enough lawmakers and secretaries of Energy came to believe that so much money and effort had been devoted to it that it was time to give it a chance. The EPA's certification of the site—though the subject of considerable debate—sufficed to them as proof of safety.

Lessons Learned from WIPP

The EPA's independent seal of approval set the single most important precedent for the regulation of a federal nuclear waste site. The era has passed in which a lone government agency can determine whether something it built to store such dangerous materials is safe. Had the concept of EPA certification been broached in the beginning, some of the controversy over the project might well have been avoided, or at least lessened.

But WIPP's history offers other lessons. From the start, support in Carlsbad that was rooted in economic anxiety gave an essential impetus to the project. The Energy Department was most successful when it capitalized on the initiative of local political and business leaders, whose unwavering backing and aggressive lobbying helped keep the project going. Obviously, then, it makes sense to locate nuclear waste sites in places where there is potential for a receptive reaction. But while any decision to site a nuclear waste facility that is based on economic and political considerations may satisfy one group of affected interests, it is unlikely to pacify the larger community that perceives itself as shouldering an unpleasant burden. Money, training, and other forms of compensation are essential. The department did little early on to ensure New Mexico would receive such benefits and paid a political price when it became dependent on congressional action to proceed. As one study of high-level nuclear waste siting noted, for an equitable outcome to occur, "construction should result in an equal change in the welfare of all parties involved (whether individuals, corporations or political jurisdictions), and this change should be beneficial."[4]

Other problems arose because the department did not clearly define its mission at the outset. Seeking as much flexibility as possible, it sent mixed messages on whether it wanted to store commercially generated as well as defense waste at WIPP. This, again, contributed to a perception that the department was only interested in forcing its will on New Mexicans, and the result was an inability to forge a consensus. British sociologist Andrew Blowers defines the formulation of a social consensus on waste disposal as "the achievement of a sufficient concurrence of view at various stages to legitimize a decision to proceed with a particular course of action." He has concluded that a consensus-based process needs guidelines reflecting not just technical but political and social components of the problem. To that end, two principles should be adopted: the early involvement of all of those who are potentially affected by a proposal, and openness and transparency throughout the policy-making process.[5] In the case of WIPP, Carlsbad's leaders did not tell anyone what they were doing in the early 1970s, allowing initial planning to take place before the public was brought into the picture. The situation created a polarized climate in which consensus was in essence beyond reach before it could be attempted.

To its credit, the Energy Department became more aware about seeking consensus. The formation of the New Mexico Environmental Evaluation Group (EEG) began a process in which the department gradually opened itself up to formal criticism. Nevertheless, the problems raised by such groups rarely forced it to bend its timetable for bringing waste. That led to accusations of political expediency, which led to more confrontation and delays. An internal 1985 report acknowledged the conflict between meeting goals and trying to involve New Mexico officials: "When the DOE schedule appears to be driving the siting program, in the absence of careful attention to emerging technical problems which might compromise the public health and safety, the state apparatus goes into high gear. . . . Unwillingness on the part of DOE to interrupt the schedule led to an increasing and perhaps unrealistic assessment of the risks involved in the project."[6] As a result, schedules should be flexible and set with an understanding that they are likely to change.

Although the evaluation group has been held up as a model watchdog of a nuclear waste project, the arrangement has had its limitations. The EEG has lacked regulatory power and concentrated on technical issues. One 1988 study funded by the Energy Department suggested a parallel "Socioeconomic Evaluation Group" be established to look at issues such as emergency response training for hospitals in communities along the waste shipping routes. The study concluded that with such a group, "attitudes and reactions by local groups—tourists, retirees and others—would have been monitored to the end of anticipating problems associated with the waste facility, as opposed to current conditions wherein the state and local communities can only react" to problems.[7] At the very least, such a group might have helped control costs. The EEG's Robert Neill has observed that the continued need for more funding has reinforced the notion among the public that waste disposal presents a formidable challenge: "People intuitively reject the argument that says we have the technology to solve the problem while we continue to request more money to study and evaluate it."[8]

Perhaps more than anything else, the battling over WIPP highlighted the complexities of science in shaping social and political decisions. The project had its genesis in scientific uncertainty: Could the United States permanently dispose of radioactive wastes through a geologic solution? Proponents used that question to justify continued research and construction efforts. They took up science in advancing the notion that WIPP, from an environmental and technological perspective, was a state-of-the art facility and should be given a chance to work. As their argument went, the plant must be carried out to its fullest conclusion—putting waste underground—to determine its success. But many technical controversies in policy making cannot be resolved simply by resorting to scientific evidence because "scientific" solutions can be subjected to social, political, or economic biases. As one political scientist has observed: "Experts can be readily, even unintentionally, caught up in the emotionally and politically polarizing atmosphere of such disputes."[9] At WIPP, because much of the scientific work came after construction began, many skeptics always assumed it was being done to reach a predetermined conclusion. University of New Mexico geologist Roger Anderson, for one, dismissed the government's approach as "mission-driven science" as opposed to an open-minded inquiry.[10]

Relying on science, then, does not always ensure consensus. By its very nature, it produces data that can be imperfect and not always easily interpreted. Scientific doubt has been and will continue to be a powerful weapon for critics of nuclear waste disposal. It was often invoked as a tactic on WIPP, and critics have continued to contend the scientific consequences of the agency's actions have not been fully proved as a way to file further litigation and attempt to influence public opinion. There may be no way to fully remedy this situation, but at the very least, the federal government should be as open as possible about revealing and discussing scientific problems. The credibility gap on WIPP widened as technical issues that caught the attention of the public and news media were raised not by Energy Department scientists but outside critics such as Anderson. Whether these technical issues were problems or not, a perception arose that the department was trying to hide or minimize them.

The 1993 Energy Department task force that examined strengthening public trust and confidence in managing nuclear wastes recognized that these issues deserved serious attention. It concluded that the agency needed to follow an elaborate series of steps when dealing with outside parties. They included seeking the "early, continuous involvement" of state and local advisory groups, carrying out agreements through an open process that had been established in advance, and making "consistent and respectful efforts" to reach out to state and community leaders and the public. Internally, the Energy task force said the department should promote high levels of professional competence through rigorous training, establish reasonable technical milestones, and reward "honest self-assessment" that enabled the agency to get ahead of problems before outsiders can discover them. "Public trust and confidence is not a luxury," it said. "DOE not only has an obligation to earn it, but it also has a compelling need to do so."[11]

Looking at High-Level Disposal

Some of these recommendations were pursued on WIPP; others were not. The question now is whether any can apply to the department's high-level-waste storage program, a more expansive and complex undertaking. Under the Nuclear Waste Policy Act of 1982, the federal government had hoped to begin loading waste into a repository by 1998, but the Energy Department does not expect to open Nevada's Yucca Mountain—still the only site being studied—until 2010 at the earliest, and that date is likely to slip. The department plans to apply for an NRC license in 2004 to build the repository. With President George W. Bush's administration placing a greater emphasis on nuclear energy as a part of what it calls a more balanced energy policy, nuclear power's proponents are under pressure to show they can do something about the waste before new nuclear plants are built. No new licenses have been issued since 1978, the year before the Three Mile Island accident.

Within the Energy Department and scientific community, WIPP has been regarded as a critical launch pad for Yucca Mountain. "A 'no' answer on WIPP would have made a 'yes' answer on Yucca infinitely more difficult," said Thomas Grumbly, the department's undersecretary and assistant secretary for waste management during the 1990s. "A 'yes' answer on WIPP keeps the options open for a 'yes' answer on Yucca."[12] Interestingly, Yucca Mountain officials have repeated some of the national security arguments made for WIPP, noting that about 10 percent of the 77,000 tons of material to be buried in Nevada is from the navy's nuclear-powered submarine reactors. "Yucca Mountain is fundamentally tied to solving the defense waste problem," said George Dials, who ran WIPP's Carlsbad office in the 1990s and later oversaw efforts in Nevada.[13]

But comparing Yucca Mountain and WIPP is difficult. In addition to its larger size and more dangerously radioactive materials, the Nevada project has the added complexity of accepting most of its waste from commercial utilities. The utilities have been clamoring for a permanent disposal site and have filed lawsuits against the government alleging that a statutory deadline to take the waste off their hands by 1998 was not met. "We can store spent nuclear fuel on-site in a very safe, efficient and cost-effective way, but that doesn't mean DOE should ignore its obligation to move the material to a permanent repository," said Steven Kraft, director of used-fuel management for the Nuclear Energy Institute, the industry's Washington lobbying arm.[14]

The National Research Council, which runs the National Academy of Sciences, has echoed its views on WIPP in endorsing the general concept of isolating highly radioactive materials in geologic formations until they have decayed to safe levels. "There is no scientific or technical reason to think that a satisfactory geological repository cannot be built," the council said.[15] As WIPP demonstrated all too vividly, however, storing nuclear waste is not just about science and technology. It is every bit as much about combating misperceptions and attempting consensus. Unlike New Mexico's state government, Nevada politicians have remained consistent in their commitment to fighting Yucca Mountain, pointing to potential volcanic activity, earthquakes, water infiltration, underground flooding, nuclear chain reactions, and fossil fuel and mineral deposits that might encourage future human intrusion. "The site itself is pretty poor from a scientific perspective, and we believe it should be disqualified for that reason," said Robert Loux, director of the Nevada Agency for Nuclear Projects, the state's anti-Yucca watchdog group.[16] Loux and others have echoed WIPP's critics of the EPA in claiming that the NRC is heavily biased in favor of licensing the Nevada project. Nevada senator Harry Reid, who became the Senate's second-most-powerful Democrat in 2001, has vowed to do everything in his power to slow or stop the project. Reid was instrumental in securing enough votes to sustain a presidential veto of legislation in 1997 that would have established a temporary above ground storage site near Yucca Mountain until the permanent facility was ready; he prevailed again in 2000 when President Bill Clinton vetoed similar legislation.

Nevada lawmakers are following in the tradition of WIPP's critics in seeking to have the courts determine Yucca Mountain's fate. They have pointed to WIPP in citing the department's inability to comply with New Mexico officials' wishes on the state-issued Resource Conservation and Recovery Act (RCRA) hazardous-waste permit. "Like the situation emerging at WIPP, if the Yucca Mountain project were ever to go forward, Nevadans should expect DOE to look for ways of cutting corners and saving money at the expense of safety-related commitments once the government has what it wants," an Agency for Nuclear Projects official said in a Nevada newspaper letter to the editor in 2000. "As with all aspects of the federal nuclear waste disposal program, a healthy skepticism is appropriate."[17]

In February 2002, Bush accepted Energy Secretary Spencer Abraham's recommendation that the science behind Yucca Mountain was sound and that work should proceed on obtaining an NRC license. But Nevada lawmakers employed many of the same arguments that WIPP's critics had used: too many scientific questions remained unanswered, transporting the waste posed an unnecessary risk, and there was little urgency to act.

The continuing stalemate between Nevada and the federal government and the likelihood of a long legal battle has led some observers to suggest that the Energy Department reconsider its plans for permanent storage at Yucca Mountain, perhaps by putting money into experimental technologies that can lessen the radioactivity of waste. It also has prompted some in New Mexico to wonder if WIPP might someday become the United States' high-level-waste site. The 1992 land withdrawal law prohibits high-level waste. But the unease reached such a level that state attorney general Patricia Madrid wrote to Senator Pete Domenici in February 2001 urging him to oppose any efforts to consider putting high-level waste at WIPP.[18]

Environmentalists, having fought so many other battles for so long, say they are prepared to fight that one if necessary. "The general perception [among nuclear supporters] is, 'Gee, New Mexico must be an easy target because we've got WIPP open,'" said Don Hancock of Albuquerque's Southwest Research and Information Center, who entered his fourth decade as the plant's leading critic. "They thought that once it was open, the opposition would all go away.

"We're trying to show that's not going to happen."[19]

Notes

Introduction

1. U.S. General Accounting Office, "Uncertainties in Opening the Waste Isolation Pilot Plant," GAO/RCED-96-146 (Washington, D.C.: GAO, July 1996), 15.

2. Gregory Benford, *Deep Time: How Humanity Communicates Across Millennia* (New York: Avon Books, 1999), 34.

3. Letter to George Dials from Dr. Claudio Pescatore on behalf of the Joint OECD Nuclear Energy, Agency/International Atomic Energy Agency, 9 April 1997, DOE Archives, Germantown, Md.

4. Associated Press, "Off-Again, On-Again WIPP Shipments Are on Again," 13 October 2001.

5. Paul Slovic, Mark Layman, and James H. Flynn, "Perceived Risk, Trust and Nuclear Waste: Lessons from Yucca Mountain," in *Public Reactions to Nuclear Waste: Citizens' Views of Repository Siting*, Riley E. Dunlap, Michael E. Kraft, and Eugene A. Rosa, eds. (Durham and London: Duke University Press, 1993), 78–81.

6. Donald L. Barlett and James B. Steele, *Forevermore: Nuclear Waste in America* (New York: W. W. Norton & Company, 1985), 48.

7. Deborah Reade, "Everything You Always Wanted to Know about WIPP" (online paper published by Citizens for Alternatives to Radioactive Dumping, 1996).

8. Michael B. Gerrard, *Whose Backyard, Whose Risk: Fear and Fairness in Toxic and Nuclear Waste Siting* (Cambridge, Mass.: MIT Press, 1994), 5.

9. Tad Bartimus and Scott McCartney, *Trinity's Children: Living Along America's Nuclear Highway* (New York: Harcourt Brace Jovanovich, 1991), 195.

10. Dixy Lee Ray and Anibal L. Taboas, "Waste Management: The Missing Link" (paper presented at Waste Management '85 conference in Tucson, Ariz., March 1985), 7–9.

11. Carl J. Mora, *Sandia and the Waste Isolation Pilot Plant 1974-1999* (draft) (Albuquerque, N.Mex.: Sandia National Laboratories, 1999), 9.

12. N. S. Nokkentved, "Buried Radioactive Waste Becomes Legacy of INEL," *Twin Falls Times-News*, 29 August 1989.

13. League of Women Voters Education Fund, *The Nuclear Waste Primer: A Handbook for Citizens* (New York: Lyons & Burford, 1993), 20–23, 36–37.

14. Barlett and Steele, *Forevermore*, 27–28.

15. William W. Hambleton, "Historical Perspectives on Radioactive Waste Management," *Kansas Water News* 23, nos. 1, 2 (1981): 4.

16. F. G. Gosling and Terrence R. Fehner, *Closing the Circle: The Department of Energy and Environmental Management 1942-1994* (draft) (Washington, D.C.: Department of Energy, History Division, March 1994), 14.

17. Arjun Makhijani, Howard Hu, and Katherine Yih, eds., *Nuclear Wastelands: A Global Guide to Nuclear Weapons Production and Its Health and Environmental Effects* (Cambridge, Mass.: MIT Press, 1995), xv.

18. Keith Rogers, "Author Details an 'Atomic West' in Lecture," *Las Vegas Review-Journal,* 21 November 1996.

19. Stanley M. Nealey and John A. Hebert, "Public Attitudes Toward Radioactive Wastes," in *Too Hot to Handle? Social and Policy Issues in the Management of Radioactive Wastes,* ed. Charles A. Walker et al. (New Haven and London: Yale University Press, 1983), 94.

20. Katherine N. Probst and Adam I. Lowe, "Cleaning Up the Nuclear Weapons Complex: Does Anybody Care?" (Washington, D.C.: Resources for the Future Center for Risk Management report, January 2000), vii.

21. "Report on Disposal of Radioactive Waste on Land" (Washington, D.C.: National Academy of Sciences/National Research Council Division of Earth Sciences, 1957), 4.

22. Barlett and Steele, *Forevermore,* 282–85; Hambleton, "Historical Perspectives," 5.

23. Gosling and Fehner, *Closing the Circle,* 15.

24. Barlett and Steele, *Forevermore,* 286–87.

25. "WIPP Provides Economic Stability for Carlsbad" (Carlsbad, N.Mex.: Department of Energy Carlsbad Area Office fact sheet, June 1999).

26. John Fleck, "Ancient Bacterium Gets New Life," *Albuquerque Journal,* 19 October 2000.

27. League of Women Voters Education Fund, *The Nuclear Waste Primer,* 36.

Chapter 1

1. Leslie Linthicum, Ian Hoffman, and Miguel Navrot, "First WIPP-Bound Load Shoves Off From LANL," *Albuquerque Journal,* 26 March 1999.

2. Department of Energy news release, Washington, D.C., 27 March 1999.

3. News release from office of Representative Joe Skeen, Washington, D.C., 26 March 1999.

4. United Press International, "AEC, Kansas Battle Over Nuclear Dump Site," *Albuquerque Journal,* 7 November 1971.

5. Michael Barone, Grant Ujifusa, and Douglas Matthews, *The Almanac of American Politics 1978* (New York: E. P. Dutton, 1978), 547.

6. Eddie Lyon interview, Carlsbad, N.Mex., August 1997.

7. *Albuquerque Journal,* "Candidates Face Issues," 8 October 1972.

8. Terry Marshall, *Carlsbad* (Carlsbad, N.Mex.: Riverside Research, 1998), 68.

9. William A. Keleher, *The Fabulous Frontier* (Albuquerque: University of New Mexico Press, 1962), 278–79.

10. Ibid., 280–81; Jack Stilton, "Promoters and Pioneers," *New Mexico,* October 1949, 24–25, 53.

11. Carl J. Mora, *Sandia and the Waste Isolation Pilot Plant 1974-1999* (draft) Sandia National Laboratories, 1999), 17–18.

12. Richard Rhodes, "Waste of the Pecos," *Playboy,* August 1979, 158; Mora, *Sandia and the Waste Isolation Pilot Plant,* 18.

13. "The 20th Century's Top Leaders," *Carlsbad Current-Argus,* "Our Town," February 1999, 32; "Gant's Phone Call Helps Mold the Future for WIPP," *Carlsbad Current-Argus,* 1 November 1992; Toni Walker, "Whitlock: A Man Who Gets Things Done," *Carlsbad Current-Argus,* 23 February 1997.

14. Lyon, interview; *Carlsbad Current-Argus,* "Gant's Phone Call."

15. *Carlsbad Current-Argus,* "Red Tape Hampers City Projects," 25 February 1979.

16. Phil Niklaus, "Waste Site Conflicts Looms," *Albuquerque Journal,* 25 June 1976.

17. Richard White, *"It's Your Misfortune and None of My Own": A New History of the American West* (Norman, Okla.: University of Oklahoma Press, 1992), 353–54.

18. Victoria Parker-Stevens, "Project Gnome: A Big Boon?" *Carlsbad Current-Argus,* "Our Town 2000," undated.

19. Chuck McCutcheon, "Carlsbad's Blast from the Past," *Albuquerque Journal,* 7 July 1991; *Annual Report to Congress of the Atomic Energy Commission* (Washington, D.C.: Atomic Energy Commission, January 1962), 208–15; Mora, *Sandia and the Waste Isolation Pilot Plant,* 32–33.

20. Louis Whitlock interview, Carlsbad, N.Mex., August 1997.

21. *Carlsbad Current-Argus,* "Red Tape Hampers City Projects."

22. Michael B. Gerrard, *Whose Backyard, Whose Risk: Fear and Fairness in Toxic and Nuclear Waste Siting* (Cambridge, Mass.: MIT Press, 1994), 111.

23. Paul Slovic, "Psychological Factors in the Perception and Acceptability of Risk: Implications for Nuclear Waste Management" (paper presented at the Conference on Public Policy Issues in Nuclear Waste Management, October 27–29, 1976).

24. *Albuquerque Journal,* "Runnels Backs Atom Waste Plant, Sees State as 'Disposal Capital,'" 25 August 1972.

25. Letter from Claude E. Wood to Joseph Gant, 24 January 1972, DOE Archives, Germantown, Md.

26. *Carlsbad Current-Argus,* "Gant's Phone Call."

27. Letter from N. William Muller to Dr. Frank Pittman, February 16, 1972, DOE Archives, Germantown, Md.; Phil Niklaus, "Failing Potash Firm Started Push for Carlsbad Disposal Site," *Albuquerque Journal,* 9 April 1978.

28. Toney Anaya interview, Santa Fe, N.Mex., August 1999.

29. Letter from James L. Harvey to Joe Gant, 15 January 1972, DOE Archives, Germantown, Md.

30. Niklaus, "Failing Potash Firm Started Push for Carlsbad Disposal Site."

31. Larry Calloway, "Cultures Color N.M. Politics," *Albuquerque Journal,* 19 September 1999.

32. Tony Hillerman, "King's Dilemma: Save Our Land? Save Our Youth?" *New Mexico,* winter 1972, 14–15.

33. Letter from Don H. Baker Jr. to Joe Gant, 7 December 1971, DOE Archives, Germantown, Md.; New Mexico Bureau of Mines and Mineral Resources, *Director's Newsletter,* September 1972.

34. U.S. Department of Energy memorandum from Owen Gormley to Dorner Schueler, 1 February 1979, DOE Archives, Germantown, Md.; Mora, *Sandia and the Waste Isolation Pilot Plant,* 13–14.

35. Gary L. Downey, "Politics and Technology in Repository Siting: Military Versus Commercial Nuclear Wastes at WIPP 1972–1985," *Technology in Society* 7 (1985): 53–54.

36. Niklaus, "Failing Potash Firm Started Push for Carlsbad Disposal Site."

37. Luther J. Carter, *Nuclear Imperatives and Public Trust* (Washington, D.C.: Resources for the Future, 1987), 177.

38. Whitlock interview, Carlsbad, N.Mex., August 1997.

39. United Press International, "Atomic Waste Disposal Site Eyed In State," *Albuquerque Journal,* 15 August 1972.

40. Draft form letter from August 1972 dealing with nuclear waste from Governor Bruce King to concerned New Mexico citizens, New Mexico State Archives, Santa Fe, N.Mex.

41. Letter from Governor Bruce King to Colonel Robert W. Rowden, 26 March 1973, New Mexico State Archives, Santa Fe, N.Mex.

42. Anaya interview, Santa Fe, N.Mex., August 1999.

43. "Statement of Walter Gerrells, Mayor, City of Carlsbad, New Mexico," House Subcommittee on Oversight and Investigations, Committee on Interior and Insular Affairs, Oversight Hearings on Nuclear Waste Isolation Pilot Plant (WIPP), 96th Congress, 1st sess., 1980.

44. Lyon, interview.

45. *Carlsbad Current-Argus,* "Mayor Supports Plan to Locate Plant Here," 15 August 1972.

46. Ned Cantwell, "People of Lyons, Kansas, Wanted AEC Plant," *Carlsbad Current-Argus,* 16 August 1972.

47. *Carlsbad Current-Argus,* "Salt Vault Plan Voiced," 13 July 1973.

48. *Radioactive Waste Repository Annual Progress Report for Period Ending September 30, 1972* December 1972. DOE Archives, Germantown, Md.

49. Mora, *Sandia and the Waste Isolation Pilot Plant,* 14.

50. Carter, *Nuclear Imperatives and Public Trust,* 180; Whitlock, interview; Toni Walker, "Ray Recalled As Early WIPP Proponent," *Carlsbad Current-Argus,* 10 January 1994.

51. Gosling and Fehner, *Closing the Circle,* 23.

52. Mora, *Sandia and the Waste Isolation Pilot Plant,* 27.

53. Downey, "Politics and Technology in Repository Siting," 55–56.

54. Whitlock, interview.

Chapter 2

1. Carl J. Mora, *Sandia and the Waste Isolation Pilot Plant 1974-1999* (draft) (Albuquerque: Sandia National Laboratories, 1999), 35.

2. Mora, *Sandia and the Waste Isolation Pilot Plant,* 35–37; Philip M. Boffey, "Radioactive Waste Site Search Gets Into Deep Water," *Science,* October 1975, 361.

3. Steve Cornett, "Geophysicist Says Carlsbad Not Ideal for Nuclear Waste," *Amarillo News-Globe,* 24 February 1973.

4. U.S. Geological Survey, "Geologic Disposal of High-Level Radioactive Wastes: Earth-Science Perspectives" (Washington, D.C.: U.S. Geological Survey, 1978), 12.

5. Associated Press, "Geologists Raise Questions on Stability of WIPP Site," *Albuquerque Journal,* 28 April 1981.

6. Mora, *Sandia and the Waste Isolation Pilot Plant,* 37; Wendell Weart interview, Albuquerque, N.Mex., August 1999.

7. Mora, *Sandia and the Waste Isolation Pilot Plant,* 38.

8. Transcript of Sandia National Laboratories interview with Wendell Weart, Albuquerque, N.Mex., October 1991.

9. Weart, interview.

10. Eddie Lyon interview, Carlsbad, N.Mex., August 1997; John German, "Sultan of Salt Becomes Third Sandia Fellow," *Sandia Lab News,* 9 May 1997.

11. Weart, interview.

12. Phil Niklaus, "Waste Site Conflict Looms," *Albuquerque Journal,* 25 June 1976.

13. Luther Carter, *Nuclear Imperatives and Public Trust* (Washington, D.C.: Resources for the Future, 1987), 181.

14. Niklaus, "Waste Site Conflict Looms."

15. Mora, *Sandia and the Waste Isolation Pilot Plant,* 49.

16. Sandia National Laboratories, *Draft Environmental Impact Statement for Waste Isolation Pilot Plant, Eddy County, New Mexico* (unissued version), April 1977, 1–17, 1–18.

17. Mora, *Sandia and the Waste Isolation Pilot Plant,* 54–55.

18. Ibid., 55; Fred C. Shapiro, *Radwaste: A Reporter's Investigation of a Growing Nuclear Menace* (New York: Random House, 1981), 115.

19. F. G. Gosling and Terrence R. Fehner, *Closing the Circle: The Department of Energy and Environmental Management 1942-1994* (draft) (Washington, D.C.: Department of Energy History Division, March 1994), 25.

20. William J. Lanouette, "The Nation's Atomic Garbage Dump?" *National Observer,* 3 February 1973.

21. *Carlsbad Current-Argus,* "Carlsbad Finds a New 'Friend,'" 4 February 1973.

22. Richard Rhodes, "Waste of the Pecos," *Playboy,* August 1979, 158.

23. Carter, *Nuclear Imperatives and Public Trust,* 179.

24. Denise Tessier, "Both Pro, Con WIPP Groups Mobilizing," *Albuquerque Journal,* 19 November 1978.

25. "Statement of Walter Gerrells, Mayor, City of Carlsbad, New Mexico," House Subcommittee on Oversight and Investigations, Committee on Interior and Insular Affairs, *Oversight Hearings on Nuclear Waste Isolation Pilot Plant (WIPP),* 96th Congress, 1st sess., 1980.

26. Ronald G. Cummings, "New Mexico Waste Isolation Pilot Plant (WIPP): An Historical Overview" (paper for the Nevada Agency for Nuclear Projects/Nuclear Waste Project Office, June 1988), DOE Archives, Germantown, Md., 27.

27. Lanouette, "The Nation's Atomic Garbage Dump?"; Peter Montague interview, Annapolis, Md., December 1999.

28. Montague, interview.

29. Michael E. Kraft, Eugene A. Rosa, and Riley E. Dunlap, "Public Opinion and Nuclear Waste Policymaking," in *Public Reactions to Nuclear Waste: Citizens' Views of Repository Siting*, Riley E. Dunlap, Michael E. Kraft, and Eugene A. Rosa, eds. (Durham and London: Duke University Press, 1993), 3–26.

30. Written testimony of Peter Montague, House Subcommittee on Environment, Committee on Science and Technology (Washington, D.C., July 18, 1978).

31. Peter Montague, "Radioactive Waste Transportation Problems and WIPP" (background paper presented to New Mexico officials, Santa Fe, N.Mex., 12 April 1978).

32. Cited in letter from Joseph M. McGough to Wendell Weart, 6 October 1981, DOE Archives, Germantown, Md.

33. Roger Anderson, "A Praedial Universe" (lecture to the Symposium for Art and Guardianship, Albuquerque, N.Mex., 19 February 1993).

34. "New Evidence of Geologic Problems at WIPP," *WIPP News*, Southwest Research and Information Center, 10 April 1978.

35. Ibid.

36. Associated Press, "Geologists Raise Questions on Stability of WIPP Site."

37. Joseph McGough, "Presentation to Governor's Task Force" (Department of Energy paper, Santa Fe, N.Mex., 19 May 1981).

38. Memo to members of the National Academy of Sciences' WIPP panel from D'Arcy Shock, 20 February 1981, DOE Archives, Germantown, Md.

39. Mora, *Sandia and the Waste Isolation Pilot Plant*, 84.

40. Ibid., 84–85.

41. Roger Anderson interview, Albuquerque, N.Mex., August 1999.

42. Louis J. Colombo, "Organizing In a 'National Sacrifice Area': The Radioactive Waste Campaign in New Mexico" (Ph.D. diss., University of Michigan, 1981), 194–98.

43. Ibid.

44. Phil Niklaus, "Study Finds Negativism Toward Site for N-Waste," *Albuquerque Journal*, 7 April 1978.

45. Colombo, "Organizing in a 'National Sacrifice Area,'" 207.

46. *Carlsbad Current-Argus*, "Skinner: While WIPP Bill Is Signed, Opposition Looms," 1 November 1992.

47. Colombo, "Organizing In a 'National Sacrifice Area,'" 363.

48. Ibid.

49. Ibid., 362–79; Eugene Ward, "Panel Members Criticize WIPP Foe," *Albuquerque Journal*, 15 July 1978; Cummings, "WIPP: An Historical Overview," 49.

50. "Statement of John Dendahl," House Subcommittee on Oversight and Investigations, Committee on Interior and Insular Affairs, *Oversight Hearings on Nuclear Waste Isolation Pilot Plant (WIPP)*, 96th Congress, 1st sess., 1980.

51. John Dendahl interview, Albuquerque, N.Mex., August 1999.

52. Colombo, "Organizing In a 'National Sacrifice Area,'" 397–98.

53. Montague interview, Annapolis, Md., December 1999.

54. Eugene A. Rosa and William R. Freudenberg, "The Historical Development of Public Reactions to Nuclear Power: Implications for Nuclear Waste Policy," in *Public Reactions to Nuclear Waste: Citizens' Views of Repository Siting*, Riley E. Dunlop, Michael E. Kraft, and Eugene A. Rosa, eds. (Durham and London: Duke University Press, 1993), 48.

55. Michael B. Gerrard, *Whose Backyard, Whose Risk: Fear and Fairness in Toxic and Nuclear Waste Siting* (Cambridge, Mass.: MIT Press, 1994), 99.

56. Don Hancock interview, Albuquerque, N.Mex., August 1997.

57. Ibid.

58. Montague interview, Albuquerque, N.Mex., December 1999.

59. Weart interview, Albuquerque, N.Mex., August 1999.

60. Associated Press, "U.S. Releases WIPP Final Impact Study," *Albuquerque Journal,* 22 October 1980.

61. Written testimony of Peter Montague on WIPP Draft Supplemental Environmental Impact Statement (Albuquerque, N.Mex., 7 June 1979).

62. Phil Niklaus, "DOE Finds Public Atop Nuclear Issue," *Albuquerque Journal,* 23 April 1978.

63. Cited in letter from Joseph M. McGough to Wendell Weart, 6 October 1981. DOE Archives, Germantown, Md.

Chapter 3

1. Alan Ehrenhalt, ed., *Politics In America 1982* (Washington: Congressional Quarterly Inc., 1981), 378.

2. Philip D. Duncan, ed., *Politics In America 1998* (Washington: Congressional Quarterly Inc., 1997), 944–45.

3. Paul R. Wieck, "WIPP Apparently a Lively Corpse," *Albuquerque Journal,* 2 May 1980.

4. Gary L. Downey, "Politics and Technology in Repository Siting: Military Versus Commercial Nuclear Wastes at WIPP 1972–1985," *Technology In Society* 7 (1985): 57.

5. "A Graveyard for Nuclear Waste," *BusinessWeek,* 28 November 1977.

6. Downey, "Politics and Technology," 58.

7. Phil Niklaus, "N-Waste Policy Shift Proposed," *Albuquerque Journal,* 16 March 1978.

8. "Governor Declines Veto Power on Carlsbad Site," *Albuquerque News,* 30 June 1977.

9. "Keep Waste Options Open," *Albuquerque Journal,* 28 June 1977.

10. Downey, "Politics and Technology," 58.

11. Paul R. Wieck, "Nuclear Waste Findings Disturbing to Delegation," *Albuquerque Journal,* Journal, 16 March 1978.

12. Phil Niklaus, "States Denied Veto Power Over Nuclear Waste Sites," *Albuquerque Journal,* 17 July 1977.

13. Luther J. Carter, *Nuclear Imperatives and Public Trust* (Washington, D.C.: Resources for the Future, 1987), 185.

14. Douglas Martin, "Dirty Debris," *Wall Street Journal,* 29 August 1978.

15. Carter, *Nuclear Imperatives and Public Trust,* 187.

16. Downey, "Politics and Technology," 59.

17. Ciel Lonon, "State Officials Blast DOE Veto Proposal," *Roswell Daily Record,* 19 October 1978.

18. Ronald G. Cummings, "New Mexico Waste Isolation Pilot Project: An Historical Overview" (paper for the Nevada Agency for Nuclear Projects/Nuclear Waste Project Office, June 1988, DOE Archives, Germantown, Md.), 5.

19. Phil Niklaus, "DOE Finds Public Atop Nuclear Issue," *Albuquerque Journal,* 28 April 1978.

20. Hazel O'Leary interview, Chevy Chase, Md., January 2000.

21. New Mexico Energy, Minerals and Natural Resources Department, "Chronology of WIPP," June 1999, online at http://www.emnrd.state.nm.us/wipp/chronolo.htm.

22. Carter, *Nuclear Imperatives and Public Trust,* 182.

23. Tim Orwig, "Evaluation Group Formed to Study WIPP Dangers," *Santa Fe New Mexican,* 24 January 1979.

24. Carl J. Mora, *Sandia and the Waste Isolation Pilot Plant 1974-1999* (draft) (Albuquerque: Sandia National Laboratories, 1999), 97.

25. Robert Neill interview, Albuquerque, N.Mex., June 1993.

26. Chuck McCutcheon, "'Low-Level' Doesn't Mean Harmless," *Albuquerque Journal,* 25 May 1989.

27. Robert H. Neill et al., "Evaluation of the Suitability of the WIPP Site," New Mexico Environmental Evaluation Group report EEG-23, Albuquerque, N.Mex., May 1983, 135.

28. Downey, "Politics and Technology," 60.

29. Ibid., 61.

30. Paul R. Wieck, "WIPP Faces Big Problems In Congress," *Albuquerque Journal,* 19 April 1979.

31. United Press International, no title, 4 May 1977.

32. Downey, "Politics and Technology," 62–63.

33. Ollie Reed Jr., "AG Calls for WIPP Halt," *Albuquerque Tribune,* 31 December 1978.

34. "What Reassurances?" *Albuquerque Journal,* 16 July 1979.

35. Bob Duke, "State Will Retain WIPP Authority," *Albuquerque Tribune,* 30 July 1979.

36. Bob Duke, "King Showdown on WIPP Marked by Diplomacy," *Albuquerque Tribune,* 26 October 1979.

37. *Congressional Record,* 9 November 1979, 31766–67.

38. Bob Duke, "WIPP a Pawn in Chess Game Between Governor, Panel Chief," *Albuquerque Tribune,* 16 November 1979.

39. Fred C. Shapiro, *Radwaste: A Reporter's Investigation of a Growing Nuclear Menace* (New York: Random House, 1981), 116.

40. Louis Whitlock interview, Carlsbad, N.Mex., August 1997.

41. Don Hancock interview, Albuquerque, N.Mex., August 1997.

42. Downey, "Politics and Technology," 63.

43. *Congressional Record,* 13 December 1979, 35801.

44. United Press International, "Carter Signs Carlsbad WIPP Bill," *Albuquerque Journal,* 1 January 1980.

45. "New Mexico Canceled as WIPP Site," *Albuquerque Tribune,* 28 January 1980.

46. "Presidential Messages," *1980 Congressional Quarterly Almanac,* (Washington, D.C.: Congressional Quarterly Press, 1981), 38-E.

47. Thomas O'Toole, "President Seeking Permanent Sites to Store Atomic Waste, Spent Fuel," *Washington Post,* 12 February 1980.

48. Steve Lambert, "WIPP Plan Cancellation Called Just 'Smoke Screen,'" *Albuquerque Tribune,* 29 January 1980.

49. Paul R. Wieck, "Carter Faces New Battle Over WIPP," *Albuquerque Journal,* 14 February 1980.

50. Bob Duke, "House Committee Breathes New Life Into WIPP Plan," *Albuquerque Tribune,* 8 May 1980.

51. "Military Waste Issue," *1980 Congressional Quarterly Almanac,* 499.

52. Associated Press, "Carter Signs Funding Bill Aiding WIPP Research," *Albuquerque Journal,* 3 October 1980.

53. Downey, "Politics and Technology," 66.

54. Carter, *Nuclear Imperatives and Public Trust,* 188.

55. Denise Tessier, "WIPP Battle Shifts to Another Agency," *Albuquerque Journal,* 11 December 1980.

56. Mora, *Sandia and the Waste Isolation Pilot Plant,* 72.

57. United Press International, no title, 27 January 1981.

58. Ibid., 4 April 1981.

59. Jeff Bingaman interview, Washington, D.C., August 1999.

60. United Press International, no title, May 14, 1981; statement of Attorney General Jeff Bingaman, New Mexico Attorney General's Office, DOE Archives, Germantown, Md., 14 May 1981.

61. United Press International, no title, June 4, 1981.

62. "DOE Makes Important Concessions to New Mexico on WIPP in Court Agreement," *Inside Energy/With Federal Lands,* 10 July 1981.

63. Bingaman interview, Washington, D.C., August 1999.

64. "Drilling Begins at Proposed Nuclear Waste Disposal Site," Associated Press, 5 July 1981.

65. *Albuquerque Journal,* 22 May 1981.

66. Bingaman, interview.

67. Ibid.

Chapter 4

1. Cecil D. Andrus and Joel Connelly, *Politics Western Style* (Seattle: Sasquatch Books, 1998), 200.

2. Rocky Barker, "Cecil Andrus Knew How To Take a Stand," *High Country News,* 20 February 1995.

3. Remarks by Cecil Andrus to the University of Denver College of Law, February 24, 1994.

4. Barker, "Cecil Andrus."

5. Andrus and Connelly, *Politics Western Style,* 201.

6. F. G. Gosling and Terrence R. Fehner, *Closing the Circle: The Department of Energy and Environmental Management 1942-94* (draft) (Washington, D.C.: Department of Energy History Division, March 1994), 29.

7. Ibid.

8. "Comprehensive Nuclear Waste Plan Enacted," *1982 Congressional Quarterly Almanac,* (Washington, D.C.: Congressional Quarterly Press, 1983), 304.

9. Donald L. Barlett and James B. Steele, *Forevermore: Nuclear Waste In America* (New York: W. W. Norton & Company, 1985), 129.

10. Milton R. Benjamin, "New Mexico Sees Its A-Waste Site as Foreboding," *Washington Post,* 27 December 1983.

11. Letter from Donald P. Hodel to Toney Anaya, 10 February 1984.

12. Toney Anaya interview, Santa Fe, N.Mex., August 1999.

13. Garrey Carruthers interview, Albuquerque, N.Mex., August 1999.

14. Jeff Barber, "New Mexicans Mull Nuclear Waste Site," *Inside Energy/With Federal Lands,* 31 August 1987.

15. William Hart, "Mystery Paper Proposes Linking N-Dump, Collider," *Dallas Morning News,* 27 September 1987.

16. Carruthers, interview.

17. Ibid.

18. "Nevada Chosen to Receive Nuclear Waste," *1987 Congressional Quarterly Almanac,* 307; Don Hancock interview, Albuquerque, N.Mex., August 1997.

19. Chuck McCutcheon, "In Nevada, Total Agreement Makes For Heated Debate," *CQ Weekly,* 17 October 1998; Luther J. Carter, *Nuclear Imperatives and Public Trust: Dealing With Radioactive Waste* (Washington, D.C.: Resources for the Future, 1987), 79.

20. "Nevada Chosen to Receive Nuclear Waste," 308.

21. Carl J. Mora, *Sandia and the Waste Isolation Pilot Plant 1974-1999* (draft) (Albuquerque: Sandia National Laboratories), 87.

22. United Press International, no title, May 5, 1983.

23. Jonathan Rauch, "Senate Budget Panel Leaders Wage War on 'Budget-Busting' Appropriations Bills," *National Journal,* 30 November 1985.

24. Mora, *Sandia and the Waste Isolation Pilot Plant,* 81; Robert Neill interview, Albuquerque, N.Mex., August 1999.

25. Deborah Reade, "Everything You Always Wanted to Know About WIPP" (on-line paper published by Citizens for Alternatives to Radioactive Dumping, 1996).

26. Mora, *Sandia and the Waste Isolation Pilot Plant* 95–96; New Mexico Energy, Minerals and Natural Resources Department, "Chronology of WIPP," June 1999, on-line at http//www.emnrd.state.nm.us/wipp/chronolo.htm.

27. Carruthers, interview.

28. Mora, *Sandia and the Waste Isolation Pilot Plant* 102–3; interviews with Robert Neill and Lokesh Chaturvedi, Albuquerque, N.Mex., August 1999.

29. New Mexico Energy, Minerals, and Natural Resources Department, "Chronology of WIPP."

30. "Status of the Department of Energy's Waste Isolation Pilot Plant," statement of Keith O. Fultz, General Accounting Office, before the Subcommittee on Environment, Energy and Natural Resources, Committee on Government Operations, House of Representatives, 13 September 1988.

31. New Mexico Energy, Minerals, and Natural Resources Department, "Chronology of WIPP."

32. "WIPP Compliance With EPA Standards," memorandum to Denise D. Fort from Gini Nelson, New Mexico Environmental Improvement Division, DOE Archives, Germantown, Md., 2 October 1986; Marilyn Haddrill, "Official Dismisses WIPP Worries," *El Paso Times*, 7 November 1986.

33. Don Hancock, "The Wasting of America: Target–New Mexico," *The Workbook*, January–March 1988, 11.

34. Alan Ehrenhalt and Philip D. Duncan, *Politics in America 1990* (Washington, D.C.: Congressional Quarterly Press, 1990), 987.

35. Karen MacPherson, "Richardson Is Bad Boy of N.M. Delegation," *Albuquerque Tribune*, 8 October 1988.

36. Bill Richardson interview, Washington, D.C., May 2000.

37. Ehrenhalt and Duncan, *Politics in America*, 984.

38. Hancock, "The Wasting of America," 10.

39. Mora, *Sandia and the Waste Isolation Pilot Plant*, 98.

40. Keith Schneider, "Leaky Mine Threatens A-Waste Storage Plan," *New York Times*, 1 February 1988.

41. Mora, *Sandia and the Waste Isolation Pilot Plant*, 105.

42. Schneider, "Leaky Mine Threatens A-Waste Storage Plan."

43. Mora, *Sandia and the Waste Isolation Pilot Plant*, 105.

44. Richard Monastersky, "Concern Over Leaks at Radwaste Site," *Science News*, 23 January 1988.

45. Elaine Hiruo, "Proposed Substitute May Close Some Gap Between House, Senate Bills on WIPP," *Nuclear Fuel*, 25 July 1988; Paul R. Wieck, "Delegation Buries WIPP Bill for Session," *Albuquerque Journal*, 4 October 1988.

46. Richardson, interview; Tony Davis, "WIPP Beyond Belief," *Albuquerque Tribune*, 13 December 1988; *Carlsbad Current-Argus*, "Withdrawal Bill Dies in Committee," 4 October 1988.

47. Davis, "WIPP Beyond Belief."

48. Mora, *Sandia and the Waste Isolation Pilot Plant*, 106.

49. Davis, "WIPP Beyond Belief."

50. Letter from Cecil Andrus to Bill Richardson, 6 July 1988.

51. Andrus remarks to University of Denver College of Law.

52. Tamara Jones, "Nuclear Refuse Piles Up, Dump Site Is Delayed," *Los Angeles Times*, 28 November 1988.

53. Andrus and Connelly, *Politics Western Style*, 201–2.

54. Carruthers, interview.

55. Chance Conner, "Nuclear Waste Mess," *USA Today*, 15 December 1988.

56. Michael D'Antonio, *Atomic Harvest: Hanford and the Lethal Toll of America's Nuclear Arsenal* (New York: Crown Publishers, 1993), 1, 2, 4, 258.

57. Connor, "Nuclear Waste Mess."

58. Joseph A. Davis, "Nuclear Weapons Woes Await Congress," *Congressional Quarterly*, 24 December 1988.

59. Gosling and Fehner, *Closing the Circle*, 52.

60. Chuck McCutcheon, "Governors, DOE Mesh on WIPP," *Albuquerque Journal*, 17 December 1988; Matthew Wald, "Three States Ask Waste Cleanup As Price of Atomic Operation," *New York Times*, 17 December 1988.

61. Ibid.

Chapter 5

1. William Lanouette, "James D. Watkins: Frustrated Admiral of Energy," *The Bulletin of the Atomic Scientists* 46, no. 1 (January/February 1990), online at http//www.bullatomsci.org/issues/1990/jf90/jf90lanouette.html.

2. F. G. Gosling and Terrence R. Fehner, *Closing the Circle: The Department of Energy and Environmental Management 1942-1994* (draft) (Washington, D.C.: Department of Energy History Division, March 1994), 55; 1988 *Congressional Quarterly Almanac* (Washington, D.C.: Congressional Quarterly Inc., 1988), 301.

3. James Watkins interview, Washington, D.C., February 2000.

4. "Watkins: Energy Secretary," 1989 *Congressional Quarterly Almanac*, (Washington, D.C.: Congressional Quarterly Inc., 1990), 692.

5. Gregg Easterbrook, "Watkins Charts New Course for Energy Department," *Albuquerque Journal*, 11 August 1991; Lanouette, "James D. Watkins: Frustrated Admiral of Energy."

6. Ibid.

7. Gosling and Fehner, *Closing the Circle*, 53.

8. Watkins, interview.

9. Joseph A. Davis, "Watkins: New Management, Tighter Controls at DOE," *Congressional Quarterly*, 25 February 1989; Keith Schneider, "Reforms Promised by Energy Nominee," *New York Times*, 23 February 1989.

10. Keith Schneider, "Wasting Away," *New York Times Magazine*, 30 August 1992.

11. Watkins, interview.

12. Leo Duffy interview, Washington, D.C., June 1999.

13. Chuck McCutcheon, "WIPP Debate Goes Beyond New Mexico," *Albuquerque Journal*, 21 May 1989.

14. Robert Neill et al., "Review of the Draft Supplemental Environmental Impact Statement for WIPP," New Mexico Environmental Evaluation Group report EEG-41, Albuquerque, N.Mex., July 1989, iv, 24.

15. Keith Schneider, "Nuclear Waste Dump Faces Another Potential Problem," *New York Times*, 3 June 1989.

16. Chuck McCutcheon, "Team Makes Sure WIPP Worth Salt," *Albuquerque Journal,* 14 October 1991.

17. Lokesh Chaturvedi and Matthew Silva, "An Evaluation of the Proposed Tests With Radioactive Waste at WIPP" (paper presented at the Third International Conference on High-Level Radioactive Waste Management, Las Vegas, Nev., 1991); Lokesh Chaturvedi, "Evaluation of the DOE Plans for Radioactive Experiments and Operational Demonstration At WIPP," New Mexico Environmental Evaluation Group report EEG-42, Albuquerque, N.Mex., September 1989, vii–viii.

18. Letter from Lokesh Chaturvedi, December 1997.

19. Duffy, interview.

20. McCutcheon, "Team Makes Sure WIPP Worth Salt."

21. "The Alar Rebellion of 1989," *Rachel's Environment & Health Weekly* #535, 27 February 1997.

22. Martha M. Hamilton, "Seeing Green in a Concrete Way," *Washington Post,* 27 April 2000.

23. "DOE to Review Safety Before Opening Site For Plutonium Waste," *Environmental Defense Fund Newsletter* XX, no. 2 (May 1989).

24. Mary Benanti, Gannett News Service, 20 March 1989.

25. Testimony of Richard Johnson, public hearing on WIPP's Supplemental Environmental Impact Statement, 17 June 1989, DOE Archives, Germantown, Md.

26. Kathy Haq, "Anti-WIPP Businesses Join Forces," *Albuquerque Journal,* 18 September 1988.

27. Chuck McCutcheon, "WIPP Foes Flog DOE at City Hearing," *Albuquerque Journal,* 14 June 1989; "DOE Broke Promises on WIPP, King Says," *Albuquerque Journal,* 15 June 1989; "Witnesses Berate DOE for WIPP," *Albuquerque Journal,* 16 June 1989; "WIPP Hearings in Santa Fe End on Emotional Note," *Albuquerque Journal,* 18 June 1989.

28. Luther Carter, "Current Controversies Over the Waste Isolation Pilot Plant," *Environment,* September 1989, 40–41.

29. McCutcheon, "WIPP Hearings in Santa Fe End on Emotional Note."

30. Alexander G. Higgins, "UN Says Effects from Chernobyl Not to Be Felt Until 2016," *Washington Post,* 27 April 2000.

31. Tony Davis, "WIPP Beyond Belief," *Albuquerque Tribune,* 13 December 1988.

32. Elouise Schumacher et al., "Westinghouse Wins the Bid: New Contractor Faces Huge Job in Running Hanford," *Seattle Times,* 13 December 1986.

33. McCutcheon, "Witnesses Berate DOE for WIPP."

34. Ibid.

35. Philip D. Duncan, ed., *Politics in America 1990* (Washington, D.C.: Congressional Quarterly Press, 1990), 1239.

36. Don Hancock interview, Albuquerque, N.Mex., August 1999.

37. Duffy, interview.

38. Transcript of the House Government Operations Committee's Subcommittee on Environmental, Energy and Natural Resources, "Review of the Status of the Waste Isolation Pilot Plant" (Washington, D.C.: U.S. Government Printing Office), 12 June 1989, 2.

39. Ibid., 173.

40. Keith Schneider, "U.S. Delays Start of Plant to Store Nuclear Wastes," *New York Times,* 14 September 1988.

41. Gosling and Fehner, *Closing the Circle,* 56–57.

42. Chuck McCutcheon, "FBI Raids Rocky Flats, Energy Department," *Albuquerque Journal,* 7 June 1989.

43. Gosling and Fehner, *Closing the Circle,* 57.

44. Chuck McCutcheon, "Embattled WIPP Won't Open in 1989," *Albuquerque Journal,* 28 June 1989.

45. Ibid.

46. Schneider, "Wasting Away."

47. Garrey Carruthers interview, Albuquerque, N.Mex., August 1999.

48. McCutcheon, "Embattled WIPP Won't Open in 1989."

49. Chuck McCutcheon and Richard Parker, "Colorado Lawmaker Calls for Early WIPP Opening," *Albuquerque Journal,* 24 October 1989.

50. Chuck McCutcheon, "DOE Idles Rocky Flats Indefinitely," *Albuquerque Journal,* 2 December 1989.

51. Gosling and Fehner, *Closing the Circle,* 64–67.

52. McCutcheon, "DOE Idles Rocky Flats Indefinitely."

Chapter 6

1. Chuck McCutcheon, "WIPP Suit Takes New Mexico Into Battle Against Giant," *Albuquerque Journal,* 12 November 1991.

2. Keith Schneider, "Wasting Away," *New York Times Magazine,* 30 August 1992.

3. Tom Udall interview, Santa Fe, N.Mex., August 1997.

4. Schneider, "Wasting Away."

5. Don Hancock interview, Albuquerque, N.Mex., August 1997.

6. John Arthur interview, Albuquerque, N.Mex., August 1997.

7. Katie Hickox, "Energy Chief Casts Doubt on Road Funds," *Santa Fe New Mexican,* 4 April 1990.

8. Ibid.

9. Chuck McCutcheon, "WIPP Opening Slim This Year, 'And Slim Just Left Town,'" *Albuquerque Journal,* 25 March 1990.

10. Garrey Carruthers interview, Albuquerque, N.Mex., August 1999.

11. New Mexico Institute for Public Policy, "Quarterly Profile of NM Citizens," fall 1991.

12. Leo Duffy interview, Washington, D.C., June 1999; Tony Davis, "Elections, Safety Fears Give Leaders WIPP-lash," *Albuquerque Tribune,* 13 September 1990.

13. Helen Gaussoin, "Jury Rules Bypass Cut Land Value, Awards Damages," *Santa Fe New Mexican,* 20 February 1991.

14. Tony Davis, "EPA's Reilly Favors Nuclear Dump at WIPP," *High Country News,* 27 August 1990.

15. Chuck McCutcheon, "New Mexico Senators Fight Early WIPP Start," *Albuquerque Journal,* 11 October 1990.

16. Karen MacPherson, "Out of the Woods," *Albuquerque Tribune*, 11 December 1992.

17. Ibid.

18. Richard Parker, "Lujan Unfettered by Old Ties on WIPP," *Albuquerque Journal*, 1 September 1991.

19. Duffy, interview.

20. John Dendahl interview, Albuquerque, N.Mex., August 1999; *Resolution No. 4 of the Committee on Interior and Insular Affairs,* 6 March 1991, DOE Archives, Germantown, Md.

21. *Albuquerque Journal,* letter from Bill Richardson, 5 June 1991.

22. *Albuquerque Journal,* letter from John Dendahl, 8 June 1991.

23. Tony Davis, "Trucking and Testing WIPP Waste," *Albuquerque Tribune*, 18 October 1991.

24. "Energy Prodded Hill on Nuclear Waste Issue," *1991 Congressional Quarterly Almanac*, (Washington, D.C.: Congressional Quarterly Inc., 1992), 224–25.

25. Chuck McCutcheon, "New Mexico Senators Introduce Bill to Allow Tests at WIPP," *Albuquerque Journal*, 3 August 1991.

26. Udall, interview.

27. Gregg Easterbrook, "Watkins Charts New Course for Energy Department," Los Angeles Times article printed in *Albuquerque Journal*, 11 August 1991.

28. Mary Benanti, "Energy Chief Says WIPP Could Open Next Week," Gannett News Service, 3 October 1991.

29. Keith Schneider, "U.S. Is Set to Store Nuclear Waste Despite New Mexico's Objections," *New York Times*, 6 October 1991.

30. Shaun McKinnon, "Lawmakers Say U.S. Seizure Portends Trouble for Nevada," *Las Vegas Review-Journal*, 8 October 1991.

31. Lindsay Lovejoy interview, Santa Fe, N.Mex., August 1997.

32. Schneider, "Wasting Away."

33. Ibid.

34. Chuck McCutcheon, "Team Makes Sure WIPP Worth Salt," *Albuquerque Journal*, 18 October 1991.

35. McCutcheon, "WIPP Suit Takes New Mexico Into Battle Against Giant."

36. Ibid.

37. Chronology prepared by New Mexico Attorney General's Office for "Radioactive and Hazardous Waste Issues in New Mexico: The Future of WIPP" (State Bar of New Mexico Conference, Albuquerque, N.Mex., 25 September 1992).

38. "Legends In the Law: A Conversation With John Garrett Penn," *District of Columbia Bar Report*, August/September 1997.

39. Richard Parker, "Udall Asks Judge to Halt WIPP Opening," *Albuquerque Journal*, 16 November 1991.

40. Chuck McCutcheon and Richard Parker, "Federal Judge Blocks WIPP Opening," *Albuquerque Journal*, 27 November 1991.

41. Ibid.

42. Andrew Taylor, "Senate Panel Clears the Way for Transfer of Waste Site," *Congressional Quarterly,* 19 October 1991.

43. Letter from Cecil Andrus to Representative George Miller, 14 February 1992, DOE Archives, Germantown, Md.

44. Katie Hickox, "Watkins Says WIPP Faces Delays With or Without Congressional Approval," States News Service, 27 March 1992; Chuck McCutcheon, "Overtime Runs Up at WIPP," *Albuquerque Journal,* 29 April 1992.

45. New Mexico attorney general's chronology for State Bar conference.

46. Duffy, interview; "WIPP Wins One, Loses One, Stays Shut," *Defense Cleanup,* 17 July 1992.

47. Transcript of Energy Department briefing by Secretary James Watkins, 29 January 1992, DOE Archives, Germantown, Md.

48. Chuck McCutcheon, "Scientists Question WIPP Test Strategies," *Albuquerque Journal,* 11 December 1991.

49. Lokesh Chaturvedi and Matthew Silva, "An Evaluation of the Proposed Tests with Radioactive Waste at WIPP," *Proceedings of the Third International Conference on High-Level Waste Management,* Las Vegas, Nev., 601.

50. Ibid., 604.

51. Chuck McCutcheon, "Panel Urges WIPP to Continue Test Plans," *Albuquerque Journal,* 12 December 1991.

52. Letter report by the Panel on the Waste Isolation Pilot Plant, Board on Radioactive Waste Management, Commission on Geosciences, Environment and Resources, National Research Council, June 1992, DOE Archives, Germantown, Md.

53. Chuck McCutcheon, "WIPP: One Very Hot Topic," *Vistas West,* November 1992, 5.

54. Chuck McCutcheon, "Academy Clarifies Stance on Tests," *Albuquerque Journal,* 20 June 1992.

55. Telephone interview with Chris Whipple, May 2000.

56. Duffy, interview.

57. James Watkins interview, Washington, D.C., May 2000.

58. "Waste Isolation Pilot Plant Land Withdrawal Act," *Congressional Record,* 21 July 1992, 18712.

59. Ibid., 18701.

60. Ibid., 18697–98.

61. Ibid., 18695.

62. Hancock, interview; analysis comparing House and Senate land withdrawal bills by Southwest Research and Information Center, Albuquerque, N.Mex., September 1992.

63. "Nuclear Waste Dump Gets Cautious Nod," *1992 Congressional Quarterly Almanac,* (Washington, D.C.: Congressional Quarterly Inc., 1993), 260–61.

64. "Gerrells Played Key Role for WIPP," *Carlsbad Current-Argus,* 1 November 1992.

65. Watkins, interview.

66. Udall, interview.

Chapter 7

1. Hazel O'Leary interview, Chevy Chase, Md., January 2000.

2. Margret Carde, "Can Bambi Ride Herd Over Godzilla? The Role of Executive Oversight in EPA's Rulemaking for the WIPP," *University of New Mexico Law School Natural Resources Journal* 36 (summer 1996).

3. Chuck McCutcheon, "O'Leary May Give Department of Energy New Look," *Albuquerque Journal*, 28 March 1993; Federal News Service, "Transcript of News Conference with President-elect Bill Clinton," 21 December 1992.

4. U.S. Department of Energy Task Force on Radioactive Waste Management, *Earning Public Trust and Confidence: Requisites for Managing Radioactive Wastes* (November 1993), v, vi, 34, 35.

5. O'Leary, interview.

6. F. G. Gosling and Terrence R. Fehner, *Closing the Circle: The Department of Energy and Environmental Management 1942-1994* (draft) (Washington, D.C.: Department of Energy, History Division, Executive Secretariat, March 1994), 117.

7. Louis Whitlock interview, Carlsbad, N.Mex., August 1997.

8. O'Leary, interview.

9. Letter from Cecil Andrus to Hazel O'Leary, 19 March 1993.

10. Tom Grumbly interview, Washington, D.C., July 1999.

11. "At WIPP, DOE May Have Put the Cart before the Horse," *Inside Energy/With Federal Lands,* 2 August 1993.

12. Chuck McCutcheon, "WIPP Tests May Exclude N-Waste," *Albuquerque Journal,* 11 March 1993.

13. Tony Davis, "About-Face on WIPP," *Albuquerque Tribune,* 26 March 1993.

14. "Council Wants Retrieval Site Before WIPP Testing Proceeds," *Weapons Complex Monitor,* 17 May 1993.

15. Carl J. Mora, *Sandia and the Waste Isolation Pilot Plant 1974-1999* (draft) (Albuquerque: Sandia National Laboratories, 1999), 146.

16. Ibid., 149.

17. Letter from Robert H. Neill to Tom Grumbly, 21 September 1993, EEG files, Albuquerque, N.Mex.

18. Task Force on Radioactive Waste Management, *Earning Public Trust and Confidence,* 34.

19. Grumbly, interview.

20. John Fleck, "DOE Drops Underground WIPP Tests," *Albuquerque Journal,* 21 October 1993.

21. "DOE Moves WIPP Test to Lab," *Defense Cleanup,* 22 October 1993.

22. Don Hancock interview, Albuquerque, N.Mex., August 1997.

23. Wendell Weart interview, Albuquerque, N.Mex., August 1999.

24. Tony Davis, "Dead WIPP Project Carries Hefty Price Tag," *Albuquerque Tribune,* 22 October 1993.

25. Amy Sue Fromer, "The Waste Isolation Pilot Plant: Perspectives Over Time 1990–96," (master's of arts thesis in geography, University of New Mexico, August 1997).

26. Grumbly, interview.

27. Gosling and Fehner, *Closing the Circle*, 135; O'Leary, interview.

28. John Fleck, "No Waste? No Problem for Booming Carlsbad," *Albuquerque Journal*, 5 March 1994.

29. John Fleck, "Overstaffed WIPP to Cut Work Force," *Albuquerque Journal*, 27 August 1994.

30. Grumbly, interview.

31. George Dials interview, Carlsbad, N.Mex., August 1997.

32. Weart, interview.

33. Keith Easthouse, "New Rules for WIPP Are More Stringent," *Santa Fe New Mexican*, 9 December 1993.

34. Dials, interview.

35. Hancock, interview.

36. Dials, interview.

37. Letter from Lindsey Lovejoy to Hazel O'Leary, 20 September 1994, Environmental Evaluation Group files, Albuquerque, N.Mex.

38. Roger Anderson interview, Albuquerque, N.Mex., August 1999.

39. Kai Erikson, "12,001 AD: Are You Listening?" *New York Times Magazine*, 6 March 1994.

40. Deborah Reade, "Everything You Always Wanted to Know about WIPP" (on-line booklet published by Citizens for Alternatives to Radioactive Dumping, 1996).

41. Weart, interview.

42. Phil Reeves, "Earthlings Keep Off," *The Independent* (London), 4 March 1994.

43. Gregory Benford, *Deep Time: How Humanity Communicates Across Millennia* (New York: Avon Books, 1999), 33.

44. Ibid., 56.

45. Keith Easthouse, "The 10,000 Year Warning," *Santa Fe New Mexican*, 13 September 1992; Jeffrey Davis, "No Trespassing. Really," *Outside*, June 1992.

46. Weart, interview.

47. Benford, Deep Time, 85.

48. Charles Pope and Angie Cannon, "Official's 'Dumb' Idea Backfires," *Philadelphia Inquirer*, 11 November 1995.

49. Alan C. Miller and Dwight Morris, "The Travels of Hazel O'Leary," *Los Angeles Times*, 10 December 1995.

50. "Nevadans Postpone Dump Site Bill," *1996 CQ Almanac* (Washington, D.C.: Congressional Quarterly Inc., 1997), 4–30.

51. "Congressional Panel Debates Fastest Way to Open WIPP," *Nuclear Waste News*, 27 July 1995.

52. Larry Craig interview, Washington, D.C., July 2000; Steve Stuebner, "Radioactive Waste Is Hot Issue In Idaho," *High Country News* 28, no. 16 (13 September 1996); Warren Cornwall, "Idaho's New Crop: Nuclear Hot Potatoes," *High Country News* 27, no. 21 (13 November 1995).

53. Grumbly, interview.

54. Reade, "Everything You Always Wanted to Know About WIPP"; New Mexico Energy, Minerals and Natural Resources chronology of WIPP events, February 1999, on-line at http//www.enmrd.state.nm.us/wipp/chronolo.htm.

55. Keith Easthouse, "Udall Blasts Domenici for Supporting WIPP," *Santa Fe New Mexican*, 3 July 1996.

56. Keith Easthouse, "Udall Turns to Appeals Court to Stall WIPP," *Santa Fe New Mexican*, 14 April 1997; New Mexico Energy, Minerals and Natural Resources chronology.

57. Carde, "Can Bambi Ride Herd Over Godzilla?" 673–75.

58. Martha Mendoza, "Seeping, Briny, Lead-Laden Water May Stall WIPP Plan," *Albuquerque Tribune*, 20 June 1996.

59. Sharyn Obsatz, "Trial Raises Doubts About WIPP Safety," *Santa Fe New Mexican*, 14 December 1994; Weart, interview; Hancock, interview.

60. National Research Council Committee on the Waste Isolation Pilot Plant, *The Waste Isolation Pilot Plant: A Potential Solution for the Disposal of Transuranic Waste* (Washington, D.C.: National Academy Press, 1996), 3.

61. News release from Senator Larry Craig and Representative Michael Crapo, Washington, D.C., 23 October 1996.

62. Mora, *Sandia and the Waste Isolation Pilot Plant*, 152–53; General Accounting Office, "Uncertainties About Opening Waste Isolation Pilot Plant," GAO/RCED-96-146 (Washington, D.C.: GAO, July 1996).

63. James Brooke, "Underground Haven or Nuclear Hazard?" *New York Times*, 6 February 1997.

64. O'Leary, interview.

65. Dan Reicher interview, Washington, D.C., August 2000.

Chapter 8

1. Federal News Service transcript of Senate Energy and Natural Resources hearing, 22 July 1998.

2. Ibid.

3. Written responses of Federico Peña to questions from the Senate Armed Services Committee, February 1997, DOE Archives, Germantown, Md.

4. Mary O'Driscoll, "DOE Officials Tell House More WIPP Delays Are Possible," *Energy Daily*, 12 February 1997.

5. Letter from 16 members of Congress to Carol Browner and Federico Peña, 17 April 1997.

6. Pete Domenici interview, Washington, D.C., July 2000.

7. Don Hancock interview, Washington, D.C., April 1997.

8. Opinion of the U.S. Circuit Court of Appeals for the District of Columbia, *State of New Mexico vs. Environmental Protection Agency and Carol M. Browner* (96–1107), 6 June 1997.

9. Frank Marcinowski interview, Washington, D.C., January 2000.

10. Thomas Grumbly interview, Washington, D.C., July 1999.

11. Marcinowski, interview.

12. Ibid.

13. "EPA Proposes to Certify WIPP, but Site Unlikely to Open Soon," *Nuclear Waste News*, 30 October 1997.

14. Wendell Weart interview, Albuquerque, N.Mex., August 1999.

15. Robert H. Neill et al., "Evaluation of the WIPP Project's Compliance with the EPA Radiation Protection Standards for Disposal of Transuranic Waste," New Mexico Environmental Evaluation Group report EEG-68, Albuquerque N.Mex., March 1998, xvii.

16. Tom Udall interview, Santa Fe, N.Mex., August 1997.

17. Mike Taugher, "WIPP Receives Federal Go-Ahead," *Albuquerque Journal*, 14 May 1998.

18. "Lessons Learned in the EPA WIPP Certification Process," executive summary of 1999 Environmental Protection Agency internal document, EPA Office of Radiation and Indoor Air, Washington, D.C.; "Evaluation of the U.S. Environmental Protection Agency's Public Outreach Program During the Certification Process at the Waste Isolation Pilot Plant in New Mexico" (study by Phoenix Environmental and EnviroIssues, Phoenix, Ariz., April 2001).

19. Chris Roberts, "Plaintiffs in New Lawsuit Hope to Delay WIPP Opening Years," Associated Press, 17 July 1998.

20. Ian Hoffman, "Delay of WIPP Urged at Hearings," *Albuquerque Journal*, 8 January 1997.

21. "Public Opinion Profile of New Mexico Citizens," University of New Mexico Institute of Public Policy 11, no. 2 (summer 1999): 5.

22. New Mexico Environment Department, paper on RCRA permit statutory and legal background prepared as testimony for draft permit hearing, November 1998, New Mexico Environment Department files, Santa Fe, N.Mex.

23. Mary Anne Sullivan interview, Washington, D.C., August 2000.

24. Secretary of Energy Advisory Board Task Force on Radioactive Waste Management, *Earning Public Trust and Confidence: Requisites for Managing Radioactive Wastes* (Washington, D.C.: Department of Energy, November 1993), 34.

25. Grumbly, interview.

26. Letter from Susan McMichael to Cooper Wayman and Gloria Barnes, 26 September 1997, New Mexico Environment Department files, Santa Fe, N.Mex.

27. New Mexico Energy, Minerals and Natural Resources chronology.

28. Mike Taugher, "Change In Law, State Stance Clear Way for WIPP," *Albuquerque Journal*, 7 November 1997; letter from Mark Weidler to John Heaton, 14 October 1997.

29. Taugher, "Change in Law."

30. Sullivan, interview.

31. Opinion of the U.S. District Court for the District of Columbia, *Environmental Defense Fund et al. vs. James D. Watkins* (91-02527 and 91-02929), 21–22.

32. Paul Detwiler interview, Washington, D.C., August 2000.

33. Telephone interview with Peter Maggiore, September 2000.

34. Letter from Peter Maggiore to Bill Richardson, 9 October 1998; testimony of Susan McMichael to New Mexico Radioactive and Hazardous Materials Committee, 5 October 1998, New Mexico Environment Department files, Santa Fe, N.Mex.

35. Letter from Louis Gallegos to Gary Falle, 5 March 1999, New Mexico Environment Department files, Santa Fe, N.Mex.

36. Letter from Gary Falle to Louis Gallegos, 12 March 1999, New Mexico Environment Department files, Santa Fe, N.Mex.

37. Transcript of testimony of John Bredehoeft at New Mexico Environment Department hearing on RCRA permit, 3 March 1999, vol. VII, 1274, DOE Archives, Germantown, Md.

38. Associated Press, no title, 17 February 1999.

39. Larry Craig interview, Washington, D.C., August 2000.

40. Bob Fick, "Kempthorne Reaffirms April 30 Deadline for Waste Shipments," Associated Press, 26 February 1999.

41. George Lobsenz, "Richardson Proposes New Nuclear Waste Storage Plan," *Energy Daily,* 26 February 1999.

42. Memorandum Order of U.S. District Judge John Garrett Penn, *State of New Mexico ex. Rel. vs. Bill Richardson, et al.,* 91-2527, 22 March 1999.

43. "Statement by Energy Secretary Richardson on Ruling by U.S. District Judge John Garrett Penn," Energy Department news release, 22 March 1999.

44. "Statement of Governor Gary Johnson," New Mexico Governor's Office news release, 24 March 1999.

45. Barbara Ferry, "WIPP Shipment to Roll This Week," *Santa Fe New Mexican,* 23 March 1999.

46. Detwiler, interview.

47. Bill Richardson interview, Washington, D.C., May 2000.

48. Hazel O'Leary interview, Chevy Chase, Md., January 2000.

49. Leo Duffy interview, Washington, D.C., July 1999.

50. Mike Taugher, "War Over WIPP Heats Up," *Albuquerque Journal,* 12 December 1999.

51. Don Hancock interview, Albuquerque, N.Mex., August 1999.

52. Weart, interview.

53. Lawrence Spohn, "Party for WIPP," *Albuquerque Tribune,* 18 April 1999.

54. Mark Warbis, "String of Broken Promises Ends as Waste Finally Leaves Idaho," Associated Press, 28 April 1999.

55. "New Mexico Withdraws Challenge to EPA Certification," *Weapons Complex Monitor,* 10 May 1999.

56. "Retired Physicist Ends Terminal Fast Against WIPP," Associated Press, 24 June 1999.

57. Mike Taugher, "WIPP Finds Plug Missing on Nuclear Waste Canister," *Albuquerque Journal,* 26 June 1999.

58. Sullivan, interview.

59. "Domenici Proposes Amendment to Restrict WIPP Permit," Associated Press, 16 October 1999.

60. Jeff Bingaman interview, Washington, D.C., November 1999.

61. Barry Massey, "DOE Uses WIPP for Warning," *Albuquerque Journal*, 9 November 1999.

62. Taugher, "War Over WIPP Heats Up."

63. Lira Behrens, "DOE, New Mexico Near Agreement on WIPP," *Inside Energy/ With Federal Lands*, 21 August 2000.

64. Ibid.

65. Hancock interview, Albuquerque, N.Mex., August 2001.

66. Southwest Research and Information Center, "Waste Isolation Pilot Plant Update/Fact Sheet by SRIC," 21 June 2001.

67. Sue Major Holmes, "Permit Modifications Denied for WIPP," *Albuquerque Journal*, 26 September 2001; "Domenici Marks WIPP Milestone, Says Facility Serves as Example in Waste Handling," news release from Senator Pete Domenici, 6 September 2001, Washington, D.C.

68. Domenici, interview.

69. Maggiore, interview.

70. Sullivan, interview.

Epilogue

1. John Fleck, "WIPP's Folly," *Albuquerque Journal*, 17 November 1993.

2. Gerald Jacob, *Site Unseen: The Politics of Siting a Nuclear Waste Repository* (Pittsburgh: University of Pittsburgh Press, 1990), 35.

3. Paul Slovic, Mark Layman, and James H. Flynn, "Perceived Risk, Trust and Nuclear Waste: Lessons from Yucca Mountain," in *Public Reactions to Nuclear Waste: Citizens' Views of Repository Siting*, Riley E. Dunlap, Michael E. Kraft, and Eugene A. Rosa, eds. (Durham and London: Duke University Press, 1993), 81.

4. James Flynn et al., *One Hundred Centuries of Solitude: Redirecting America's High-Level Nuclear Waste Policy* (Boulder: Westview Press, 1995), 34.

5. Andrew Blowers, "Prospects for Nuclear Waste Cleanup: Is Consensus Achievable?" (paper presented at the Nuclear Free Local Authorities meeting, Rotherham, England, 28 October 1999).

6. "WIPP Case Study: An Analysis of the Open Literature Concerning WIPP-Associated Federal-State Interactions" (draft paper dated 26 July 1985, on file at U.S. Department of Energy Archives, Germantown, Md.), 19.

7. Ronald G. Cummings, "New Mexico Waste Isolation Pilot Plant: An Historical Overview," study for the Nevada Agency for Nuclear Projects, June 1988, DOE Archives, Germantown, Md., 45–46.

8. Robert H. Neill, "Understanding Perception of Risks from Nuclear Waste Disposal" (paper presented at Safewaste '93 conference, Avignon, France, 13–18 June 1993).

9. Walter A. Rosenbaum, *Environmental Politics and Policy* (Washington, D.C.: Congressional Quarterly Inc., 1995), 175.

10. Roger Anderson interview, Albuquerque, N.Mex., August 1999.

11. U.S. Department of Energy Task Force on Radioactive Waste Management report, *Earning Public Trust and ConWdence: Requisites for Managing Radioactive Wastes* (November 1993), vi, vii, 21.

12. Thomas Grumbly interview, Washington, D.C., July 2000.

13. George Dials interview, Las Vegas, Nev., August 1999.

14. Brian Hansen, "Nuclear Waste," *CQ Researcher,* 8 June 2001, 492.

15. Congressional Research Service Issue Brief IB92059, "Civilian Nuclear Waste Disposal," 12 June 2001, 4.

16. Hansen, "Nuclear Waste," 493.

17. Joseph C. Strolin, "Downside of WIPP," *Pahrump Valley Times,* 15 November 2000.

18. Letter from Patricia A. Madrid to Senator Pete V. Domenici, 1 February 2001.

19. Don Hancock interview, Albuquerque, N.Mex., August 2001.

Index

RECEIVED

DEC 17 2002

ENGINEERING LIBRARY